/ SEX IN CITY PLANTS,
ANIMALS, FUNGI,
AND MORE /

/ SEX IN CITY PLANTS, ANIMALS, FUNGI, AND MORE /

A GUIDE TO REPRODUCTIVE DIVERSITY

Kenneth D. Frank

Foreword by Jonathan Silvertown

Columbia University Press
New York

Columbia University Press
Publishers Since 1893
New York   Chichester, West Sussex
cup.columbia.edu

Library of Congress Cataloging-in-Publication Data

Names: Frank, Kenneth D., author. | Silvertown, Jonathan W., writer of
   foreword.
Title: Sex in city plants, animals, and other forms of life : a guide to
   urban reproductive diversity / Kenneth D. Frank ; foreword by Jonathan
   Silvertown.
Description: New York : Columbia University Press, [2022] | Includes
   bibliographical references and index.
Identifiers: LCCN 2021044525 (print) | LCCN 2021044526 (ebook)
   | ISBN 9780231206068 (hardback) | ISBN 9780231206075 (trade
   paperback) | ISBN 9780231556309 (ebook)
Subjects: LCSH: Sex (Biology) | Insects—Sexual behavior. | Plants, Sex
   in. | Sexual behavior in animals. | Urban ecology (Biology) | Biodiversity.
Classification: LCC QH481 .F73 2022 (print) | LCC QH481 (ebook) |
   DDC 573.6/374—dc23/eng/20211029
LC record available at https://lccn.loc.gov/2021044525
LC ebook record available at https://lccn.loc.gov/2021044526

Photo on title page and front cover: Courtship in mourning doves.

Book designed by Meena Mehta | www.TwoMs.com

/ CONTENTS /

/ FOREWORD /

The subject has been celebrated in song—"They say in Boston even beans do it" (Cole Porter, "Let's Do It")—but this is the first-ever book dedicated to sex in city plants and other wildlife. As well appreciated by Cole Porter, but perhaps by few other citizens, sex is a universal, and in the city it's all around us, though largely unnoticed. And increasingly universal are the species that are found in cities. Urban environments everywhere tend to share a set of common plant and animal species, so many of those found in this book will live on a street near you, wherever that is.

Each species' account in this book is conveniently bite-sized, engagingly written and beautifully illustrated with the author's own photographs. This is scientific natural history, so along with the wonderful facts, there are questions, hypotheses, and explanations. For example, why does ragweed produce such copious pollen? Its purpose is not to plague allergy sufferers! Each tiny ragweed flower contains only one ovule (egg) that can be fertilized by a single pollen grain, and yet each male flower produces a profusion of pollen grains. The explanation rests on the fact that these grains can only fertilize the ovules of other ragweed plants. Because of this, pollen grains from different plants compete for the limited number of ovules that they can reach on the wind. Plants that produce the most pollen win the aeolian lottery for a mate, which over evolutionary time has driven pollen production ever upward, till the air is filled with allergenic dust. If ragweed pollen forces you to stay indoors, you could do a lot worse than spend the time with this book—though reading it will certainly make you want to get out and explore the biodiversity of your city from the new perspective it offers.

Jonathan Silvertown

Threadstalk speedwell (*Veronica filiformis*).
Each flower contains both male and female
parts, but none of these flowers will make
viable seed. (See page 32.)

/ PREFACE /

Downtown Philadelphia.

City plants grow spontaneously in sidewalk cracks surrounded by buildings and streets. Small invertebrates reside in walled courtyards. How do members of apparently isolated urban populations find mates? How do they avoid inbreeding?

Plants and animals that inhabit cities are often common and widespread outside of cities. A rich literature illuminates their systems for mating. However, this literature is scattered, and it typically focuses on locations that are not urban.

This book explores the natural history of sex, from bacteria and fungi to plants and animals, excluding human beings. It focuses on cities. It examines sex as a state of being, as in male or female, and as a process, as in mating. An introductory section depicts the city as both facilitator and inhibitor of mating. Chapters that follow contain over a hundred illustrated essays on mating systems and sexual behavior. Most essays are limited to a single page and highlight one reproductive trait of one urban species or group. A concluding essay considers urban reproductive diversity as a whole.

Species that the chapters examine meet three criteria: (1) they inhabit Philadelphia; (2) published studies illuminate their reproductive biology; and (3) they illustrate reproductive diversity. Most of the species reside downtown within a walk from my home, and many occur in cities around the world. None is limited to Philadelphia or the Mid-Atlantic region. Roads, buildings, and other built structures fragment their habitat. Cited literature is global.

Compared to other forms of biodiversity, reproductive diversity is largely hidden, yet it rivals morphological diversity in beauty and scale. This book explores sex in the city as a pathway to appreciation of biodiversity. It presents many unanswered questions and hypotheses. I hope they inspire readers to seek answers.

Bisexual silver maple (*Acer saccharinum*). Functionally male flowers (top right) are blooming close to but separate from functionally female flowers (bottom left). The functionally female flowers have nonfunctioning male parts (stamens). (See page 35.)

/ ACKNOWLEDGMENTS /

The research of many investigators inspired this book. I am especially grateful to those with whom I have had personal correspondence and discussions. They include Jacek Bartlewicz, Gregory W. Cowper, Peter K. Ducey, Glavis B. Edwards, Dina M. Fonseca, Amanda Gorton, Anne Harvey, David A. Hewitt, Michael Z. Levy, Sara M. Lewis, Kevin J. McGraw, Dennis Paulson, Joseph A. Perillo Jr., Nate Rice, Alfred Ernest Schuyler, Patrick G. Meirmans, Richard Shine, Douglas B. Sponsler, Harold B. White, and Helen J. Young.

Susan Alix Williams in Rowe, Massachusetts, identified the tree mosses presented in this book. Through correspondence, Susan Munch at Albright College helped me identify hemispheric liverwort.

Residents of metropolitan Philadelphia helped me explore local natural history. They include Edward Barnard, Anne Sarah Bekker, Bernard Brown, Michael Carpenter, Louis D. Giosso, Mandy Katz, June Lauer, Scott McConnell, Janet Novak, Heather Rinehart, and Paul Vassallo.

Many photographers contributed illustrative images. I have had the pleasure of corresponding or meeting with four: Mary Anne Borge, Nigel D. F. Grindley, Frank Miles, and Leo Sheng. I have listed all the photographers on page 162. All photographs presented without attribution are mine. I took all these photos in Philadelphia except where captions note otherwise. Some of my photos appear in an earlier work of mine, *Ecology of Center City, Philadelphia*.

Susan E. Frank, Robert E. Frank, and Jean Morse guided me through this book's early conception. Beth Desautels and Diane Fredrick edited this book in its early stages, and Leslie Kriesel at Columbia University Press edited its final version. Miranda Martin of Columbia University Press and Carolyn Beans helped me shape both the content and presentation. Meena Mehta created this book's handsome design.

Female umbrella liverwort (*Marchantia polymorpha*) with spore capsules in her "umbrellas." (See page 3.)

/ INTRODUCTION /

How do wild plants and animals in the city find mates? To explore this question, I invite you to join me on a stroll around downtown Philadelphia.

Our walk starts along a block of rowhouses. We chance upon a clump of annual bluegrass (*Poa annua*) in a sidewalk crack. The plant is ankle high and about as wide as my shoe. Intermingled with its deep green blades are minute greenish-white flowers. This species inhabits almost all major cities worldwide. It occurs in more cities than does any other flowering plant.[1] This particular clump faces a common urban problem: scarcity of nearby prospective mates.

Pavement and buildings in cities fragment populations of wild plants and animals. Denied mates, solitary individuals may fail to reproduce. Small, isolated urban populations may inbreed and lose genetic diversity. They may weaken and go locally extinct.[2]

Annual bluegrass can self-fertilize and cross-fertilize. Wind pollinates it.[3] Self-fertilization ensures that our clump of annual bluegrass will produce seed regardless of its ability to secure a mate. On the other hand, cross-fertilization would potentially produce offspring with more genetic diversity and more vigor, and it would avoid inbreeding.[4]

Sidewalk Cracks

As our stroll continues, we find other solitary clumps of annual bluegrass growing in cracks in this sidewalk. The smooth, flat surface of the sidewalk eases dispersal of grass seed by wind and water. The cracks trap the seeds along with organic debris. They act as soil composters and plant nurseries. They protect seedlings from trampling.

Farther down the block, clumps of blooming annual bluegrass in sidewalk cracks coalesce into linear arrays consisting of many plants. Our sidewalk brings clumps of annual bluegrass together even though it also disperses them. We suspect that this sidewalk helps flowering annual bluegrass mix and mate, increasing chances for cross-pollination.

Walking along the sidewalk, we contemplate our roles as seed dispersers. The botanist Henry Clifford recovered viable seeds from footwear of students returning to England from a trip to Ireland. From these seeds he grew 42 species, including annual bluegrass.[5] As our shoes tread the pavement, we disperse seed of annual bluegrass.

We track these seeds onto the street, where cars pick them up. The botanist Wolfgang Schmidt collected seeds from wheels and mud guards of one car driven for one season in the surroundings of Göttingen, Germany. From these seeds he grew almost 4,000 seedlings and over 100 species, including annual bluegrass. He concluded that vehicular dispersal of seed is the source of most roadside flora.[6] Applying these findings to our stroll in the city, we suspect that our foot traffic, combined with vehicular traffic, promotes genetic mixing over a wide area. In the city, annual bluegrass has incorporated us into its mating system.

Courtyard Gardens

Now we contemplate a courtyard garden behind a rowhouse. A brick wall and the rear of the house completely enclose the courtyard. The soil organisms within this garden are trapped within the courtyard. Their potential for outbreeding would appear at first glance precariously low. But think of our own gardens. Their soil contains abundant organisms: earthworms, snails, slugs, millipedes, centipedes, pill bugs, spiders, springtails, proturans, and diverse insects.[7] Every time we plant a shrub or tree with a root ball, we incidentally introduce into the soil new invertebrates.[8] For soil animals in isolated courtyard gardens, people like us facilitate outcrossing.

Door Lamps, Glass Façades, and Highways

A few doors down from this courtyard garden a common looper moth (*Autographa precationis*) rests on a wall beside a door lamp. Here, we consider another barrier to mating in the city: light pollution. Nocturnal light from the door lamp may interrupt mating and pollination by this moth. Nocturnal light pollution compounds polarized light pollution, which occurs during the day. Polarization of light reflected off shiny surfaces like glass façades confuses flying aquatic insects that normally orient using polarization of light reflected off water. These insects mature in water but mate out of water.

Highway traffic generates acoustic noise pollution. Acoustic noise masks songs of birds, frogs, and insects. Water pollution disturbs sexual differentiation in fish. It causes intersexes—fish with male gonads containing oocytes (cells capable of forming eggs). Sexual effects of pollution are not always negative. Door lamps promote mating of spiders that feed on insects attracted to light.

Urban Canyons

Our stroll about the city reaches blocks lined with the tallest buildings. For most of the day these urban canyons shroud streets and walkways in shadows. They shut out most visible forms of life other than people and ornamental plantings. Here we find no plants colonizing pavement cracks. These blocks demonstrate how pavement and buildings act as barriers in ways that are not specific to sex. For most wild plants and animals, they block basic functions required to sustain life.

While these barriers are formidable, organisms do meet and mate here, both indoors and outdoors. They include the oriental cockroach and bedbug. Jason Munshi-South, an urban evolutionary biologist at Fordham University in New York City, and colleagues, have used molecular genetics to track dispersal of populations of wild animals in Manhattan and other parts of New York City. They discovered that several species of mammals disperse within the city. They include coyotes, Norway rats, and white-footed mice.[9] By contrast, they found that populations of the urban dusky salamander in Manhattan were genetically isolated; they were confined within two tiny aquatic seeps in a degraded urban forest.[10]

Ports, Railroads, and Interstates

We enter a skyscraper (One Liberty Place) to visit its observation deck overlooking the city. Looking out from the 57th floor, we see bridges spanning rivers; barges and freighters; ports linked to networks of interstates and railroads; and air traffic. As hubs of commerce and transportation, cities historically have catalyzed genetic mixing. In recent centuries they have done so on an industrial global scale. In the nineteenth century, sailing ships arriving with empty holds dumped ballast mixed with plant material in port cities. By 1880, ballast plants around New York City encompassed over 350 species.[11] Here genetic varieties from foreign ports around the world could meet and mate.

Hulls and cargo of ships have also unintentionally transported organisms.[12] In contrast to sailing ships, modern ships discharge ballast in the form of water. They introduce marine, brackish, and freshwater organisms around port cities.[13] The steady influx of introduced species worldwide testifies to the power of trade and shipping to disperse organisms that then mix and mate. Introductions are often intentional, such as those involving ornamental plants and fish.

Lawns and Vacant Lots

Back on the sidewalk, we walk out of the urban canyons and view municipal open space covered with turf. When not treated with herbicide, lawns support outcrossing in wild plants, as in the case of white clover. Vacant lots are scarce in this prosperous neighborhood compared to neighborhoods remote from downtown. Vacant lots function as refuges for herbaceous plants too big and unsightly for turf and sidewalk cracks. Loss of vacant lots downtown reduces genetic stock available for outcrossing of wild plants.

Green Verges and a River

We follow the sidewalk to a river trail. It runs between railroad tracks and the Schuylkill riverbank. Along this trail and tracks are green verges that connect semiwild urban green spaces. Similar verges along roads link urban forests to naturalized settings in parks and cemeteries. The river itself acts as an aquatic corridor connecting wetland plants and animals for miles upstream and downstream. A fish ladder helps migrating American shad surmount a dam.[14]

The City

The city exposes mating systems to a patchwork of habitats, each with opportunities and challenges. We have viewed a wide range of habitats, from pavement cracks to urban canyons. Most are integral to cityscapes globally. They provide places for reproductive diversity in cities around the world.

This book focuses on reproductive diversity in one city. While some of the species it presents may not occur in your region, the basic kinds of mating systems that the species represent occur worldwide and are taxonomically widespread. This book may serve as a reference for reproductive diversity in cities generally. I hope it encourages you to explore urban reproductive diversity close to home, wherever you may live. We conclude our journey with a photographic look at cities around the world.

Urban canyon, Seventh Avenue, New York City. Suppression of wild plants and animals here is not limited to sexual reproduction but is general to most functions essential for life. (See page xvi.)

Times Square 2, by Casper Moller (CC BY 2.0). Image exposure adjusted. URLs on page 162.

Annual bluegrass (*Poa annua*) in bloom, Philadelphia. The downspout has watered the plant and will soon disperse its seeds. Pavement cracks will trap many of these seeds. The cracks will protect and nurture seedlings. (See page xv.)

Vacant lot, Nazareth, Israel. Vacant lots in cities are refuges for birds and bees.[15] They are also repositories for genetic stock of wild urban plants. (See pages 17 and 26.)

Vacant Lot with Wildflowers—Nazareth—Israel, by Adam Jones (CC BY 2.0). Image cropped. URLs on page 162.

Mission Dolores Park, San Francisco. In the absence of herbicides, lawns support reproduction of wild plants, as illustrated by the carpet of tiny white flowers in the foreground. Lawns provide long-lasting, expansive habitat, encouraging cross-pollination. (See page 27.)

Memorial Day 2020—San Francisco Under Quarantine, by Christopher Michel (CC BY 2.0). Image cropped. URLs on page 162.

Powązki Cemetery, Warsaw, Poland. Cemeteries and churchyards provide stable, long-term habitat for breeding populations of wild urban plants and animals, but intensive management can degrade their biodiversity.[16] (See pages 61 and 119.)

Muslim Cemetery (Tatar) Powązki, by Jolanta Dyr (CC BY-SA 3.0). Image cropped. URLs on page 162.

Valley Parkway, Mill Stream Run Reservation, Cleveland Metroparks, Ohio. Roads disperse seeds and spores, supporting genetic mixing;[17] but they also divide and degrade habitat.[18] (See pages 86 and 90.)

Low Water Ford of Valley Parkway…, by Chris Light (CC BY-SA 4.0). Image cropped and exposure adjusted. URLs on page 162.

Jackhammer, Washington, D.C. Acoustic noise masks mating calls emitted by birds, frogs, and insects. (See page 111.)

Photo by Erik Calonius

The Seine, Paris. Pollution and damming of urban rivers interfere with migration of fish to spawning grounds. On the Seine these reproductive barriers have recently moderated, and populations of migratory fish have partly rebounded.[21] (See page 91.)

Seine @ Paris, by Guilhem Vellut (CC BY 2.0). Image cropped. URLs on page 162.

/ NONFLOWERING PLANTS /

Ferns and moss on an old masonry retaining wall downtown.

Ferns and moss produce neither flowers nor pollen. They begin the process of fertilization by release of sperm. Fertilization takes place close to the ground (or wall) and leads to growth of spore-producing structures. In the photo, these structures include the tall plants that we recognize as ferns. Some of the nonflowering plants shown in this photo produce spores without fertilization. Others self-fertilize. Still others may propagate without making spores. This chapter explores all these reproductive pathways.

Spores: dwarf bristle-moss (*Orthotrichum pumilum*)

Self-fertilization of dwarf bristle-moss produces spores that develop into bisexual plants.

Street tree hosting lichens and mosses, including dwarf bristle-moss.

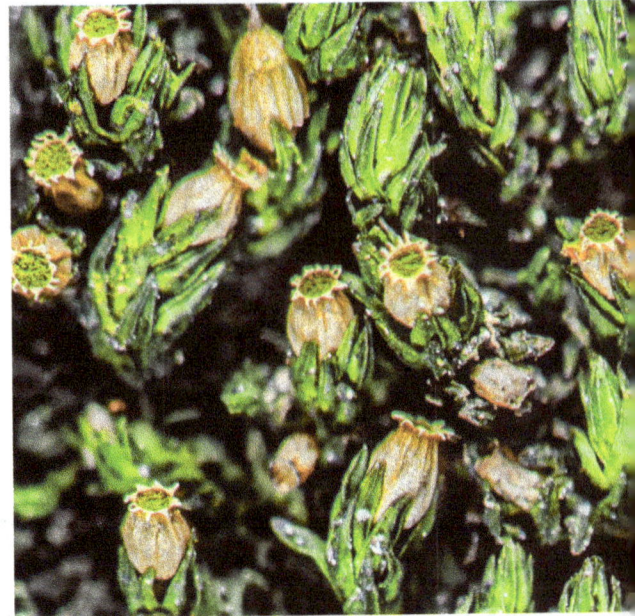

Spore capsules of dwarf bristle-moss on a street tree in downtown Philadelphia.

Dwarf bristle-moss grows on the bark of street trees in cities of Europe[1] and North America. It shares a reproductive problem common to all tree mosses: How does a moss anchored to a tree find a mate? The question applies especially in cities, where trees are spaced far apart.

In this moss, male sex organs and female sex organs are positioned on the same plant. By enabling self-fertilization, this bisexual arrangement provides assurance of fertilization and production of spores. However, male and female sex organs are spaced apart,[2] so fertilization by another plant is also possible. Cross-fertilization offers a chance for genetic mixing and new, vigorous genetic combinations.

Spores may contribute to this genetic mixing. A spore of dwarf bristle-moss consists of only one cell measuring 13 to 17 thousandths of a millimeter.[3] A colony of moss may discharge more than a hundred million spores.[4] Wind and road traffic can disperse them,[5] potentially mixing together spores from different plants and even different colonies.

In Philadelphia, dwarf bristle-moss on street trees produces abundant minute spore capsules easy to detect with the naked eye. Proliferation of this moss in the heart of the city may be credited at least in part to a mating system that allows a solitary plant to make spores with or without a sexual partner.

Cross-Fertilization: umbrella liverwort (*Marchantia polymorpha*)

Male and female sex organs of umbrella liverwort reside on separate plants. Sperm travel in water.

Male umbrella liverwort with sperm platforms.

Female with spore capsules (shown also on page xiv).

Umbrella liverwort occurs worldwide. In contrast to moss, it grows as a mostly flat, lobed green disk, or thallus, flush with the ground or substrate. It lacks leaves. Male umbrella liverworts erect star-shaped sperm platforms raised on stalks. Rain disperses sperm from these platforms. After fertilization, female umbrella liverworts produce spores. They enclose their spores in capsules attached to green umbrellas elevated on stalks. Drooping appendages radiating from the center of these umbrellas create a striking profile.[1]

Silvia Pressel and Jeffrey Duckett, bryologists at the Natural History Museum in London, investigated fertilization of umbrella liverwort that had colonized flat or undulating terrain after fire in an English nature reserve. Male plants fertilized 100 percent of female plants at a distance up to 19 meters. Active and passive transport of sperm swimming in water contributed to this success. The researchers calculated that a single episode of flooding induced a male plant to release sperm in quantities exceeding 50 million. They noted 10 such floods in the summer during the liverwort's sexual reproductive period.[2]

In downtown Philadelphia, umbrella liverwort propagates in brick pavement cracks. In rain, these cracks fill up with water. I suspect this water channels sperm to egg cells of female plants growing in the cracks. In effect, rain may transform these brick sidewalk cracks into aqueous mating corridors.

Colony of male and female liverwort in cracks between brick pavers.

Vegetative Reproduction: umbrella liverwort (*Marchantia polymorpha*)

Each sex of umbrella liverwort can propagate vegetatively independent of the other.

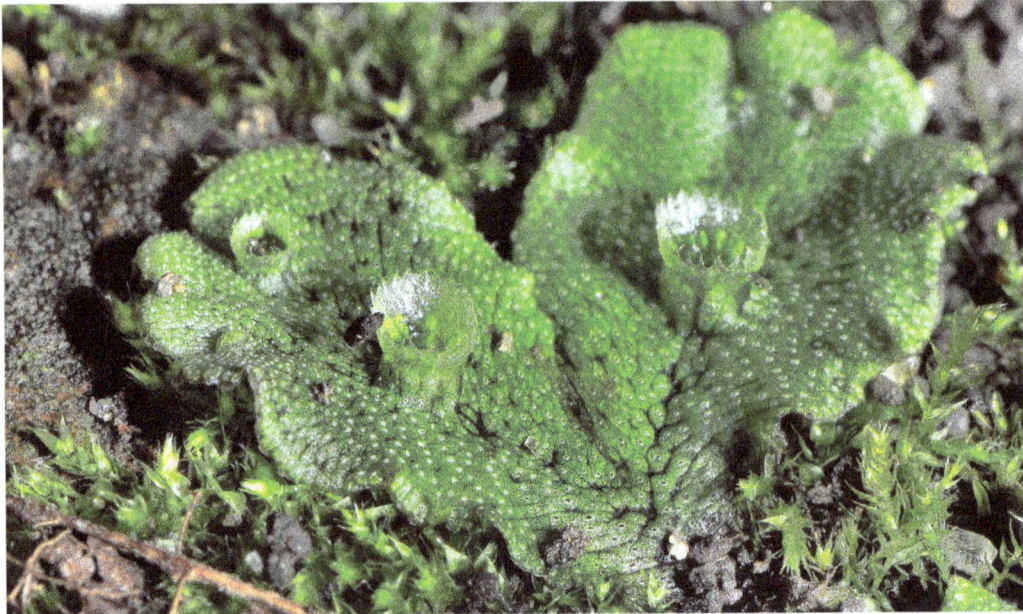

Splash cups containing green vegetative propagules (gemmae).

Umbrella liverwort has an alternative reproductive system. This system features splash cups, as shown in the photo. The splash cups contain buoyant green vegetative propagules (gemmae) dispersed by rain.[1] After rain scatters these propagules, they germinate and form new plants.[2] Both sexes make flash cups, enabling each sex to propagate on its own.

The two reproductive systems are complementary. In studies, younger plants populated colonies' outer margins and made splash cups; older plants dominated colonies' centers and made sexual structures.[3] Reproduction switched from vegetative to sexual during the second year of life.[4]

By contrast, long-distance dispersal of solitary spores may produce umbrella liverwort plants sexually isolated from all potential mates. Such plants make sterile colonies consisting of vegetative clones all of the same sex. They make no spores. Urban habitats isolated by buildings and roads might be expected to favor such colonies. Sterile female colonies have been noted in London.[5] However, colonies that I have examined in downtown Philadelphia typically have both sexes; they reproduce sexually, erecting umbrellas with spore capsules.

Liverworts and mosses may stop reproducing sexually. Over 90 percent of colonies of a species of liverwort (*Marchantia inflexa*) studied along a river in a protected watershed in Trinidad expressed no sex or only one sex.[6] Another liverwort (*Acrobolbus ciliatus*) is known as only female in the southern Appalachians and as only male in Japan.[7] Marble screw-moss presented on page 7 expresses no sex in Philadelphia.

Self-Fertilization: hemispheric liverwort (*Reboulia hemisphaerica*)

Hemispheric liverwort positions male and female sex organs on the same plant but varies the distance between them.

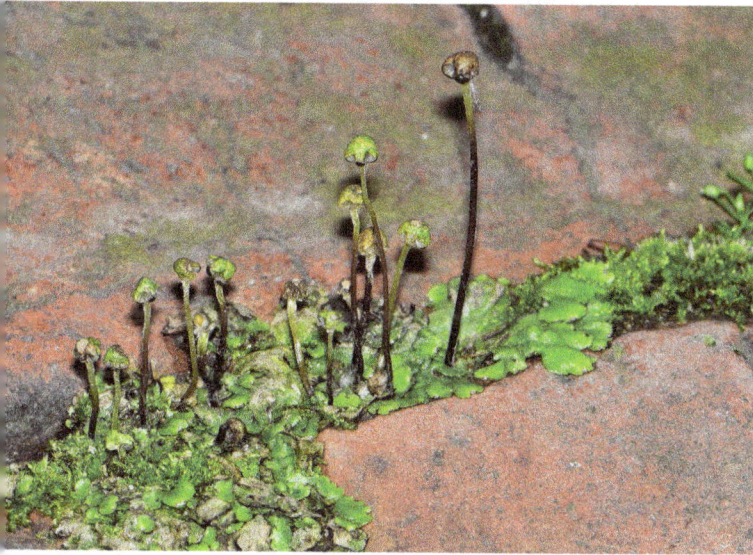

Stalked spore capsules rising from pavement crack.

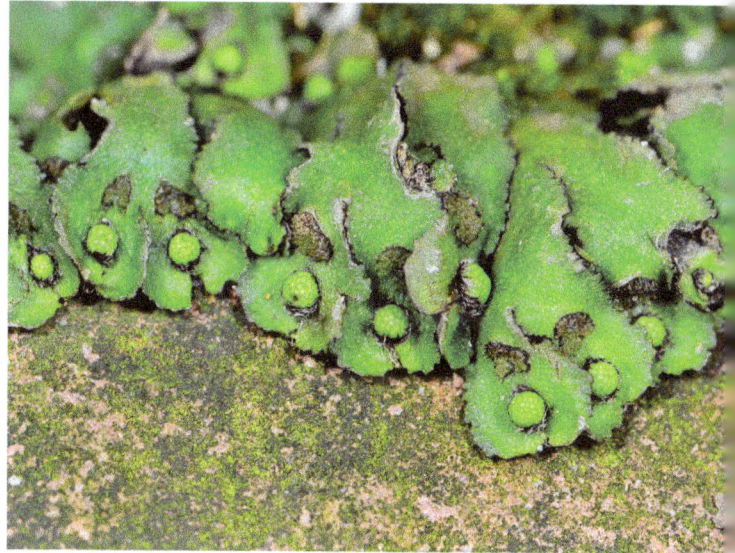

Male and female sex organs on same plant.

Hemispheric liverwort occurs on all continents except Antarctica. It grows as a branching, flat, green ribbon flush with the ground. Typically, one male sex organ and one female sex organ arise near the end of a branch. The gap between them varies.[1] On the one hand, occasionally the end of a branch has only one sex organ, allowing cross-fertilization.[2] On the other hand, some male and female sex organs of hemispheric liverwort that I have examined in Philadelphia were positioned so close together that they appeared to touch, favoring self-fertilization. Like umbrella liverwort, this liverwort produces spores enclosed in capsules elevated on tall stalks. Dispersal of these spores may facilitate genetic mixing.

Self-fertilization in hemispheric liverwort results in production of spores that are genetically identical.[3] By contrast, self-fertilization in flowering plants such as chickweed (page 10) results in genetic recombination producing seeds that are genetically different.

I suspect this liverwort's system of fertilization enables cross-fertilization opportunistically. Storm water flowing over hemispheric liverwort potentially mixes sperm with eggs over a wide area, as in umbrella liverwort. Like umbrella liverwort, hemispheric liverwort thrives in brick pavement cracks. Pavement cracks may function as aqueous mating corridors for both species.

Male sex organ on left, female on right.

Spores Without Fertilization: purple cliffbrake fern (*Pellaea atropurpurea*)

Purple cliffbrake fern is a nonflowering plant that makes spores without fertilization.

Historic penitentiary wall, habitat of ferns.

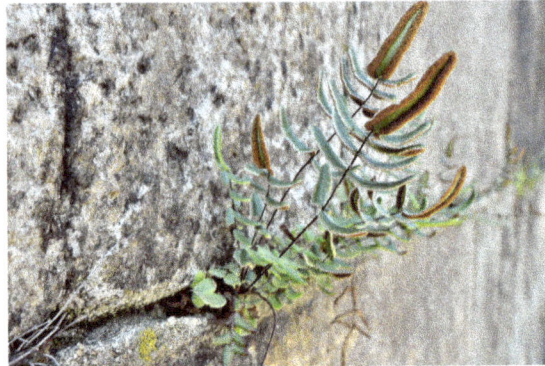
Purple cliffbrake fern with brown spores.

Like mosses and liverworts, ferns are nonflowering plants that typically require fertilization to make spores; and they need water as a medium for movement of sperm.[1] Purple cliffbrake fern is an exception. It produces spores without fertilization.[2] Embryos emerge from somatic (nongerm) cells.[3] The reproductive system of purple cliffbrake fern is well adapted to the city. A single spore carried by wind to an isolated fragment of dry habitat can establish a colony.

Ferns that reproduce like purple cliffbrake were until recently presumed to be clones. However, in 2003 Hiroshi Ishikawa, molecular biologist at Chiba University, Japan, and colleagues, showed that a Japanese species of such a fern produced offspring that were genetically diverse.[4] Thirty years earlier a study of chromosomes of ferns including purple cliffbrake had predicted such genetic diversity but did not attempt to detect it. It correctly envisioned a system of genetic recombination that operated without fertilization.[5]

In Philadelphia, purple cliffbrake fern establishes colonies on dry stone walls where it has little competition. However, another fern, ebony spleenwort (*Asplenium platyneuron*), may accompany it. Ebony spleenwort can propagate by vegetative propagules and by self-fertilization.[6] Like purple cliffbrake, it has a reproductive system that circumvents dependence on water and mates.

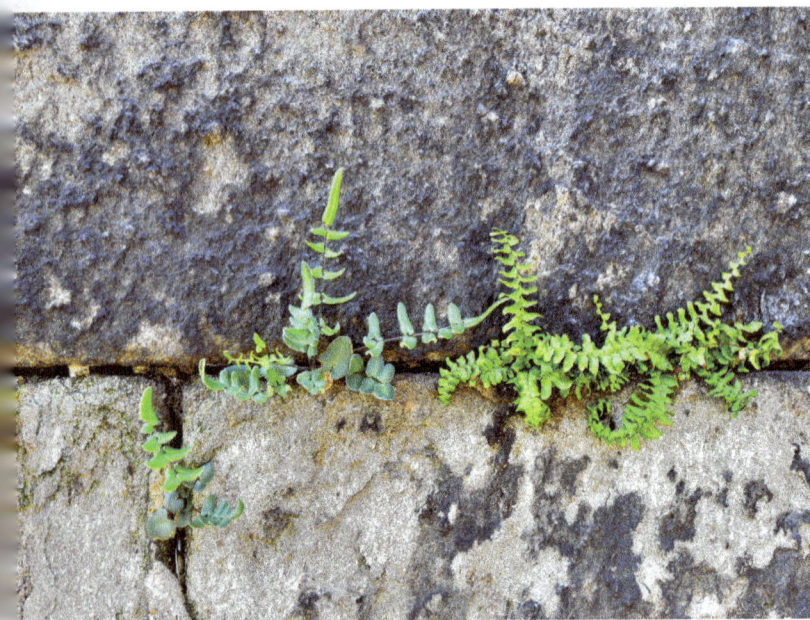
Purple cliffbrake fern (left) with ebony spleenwort fern (right).

Chapter 1 // Nonflowering Plants

Lost Sex: marble screw-moss (*Syntrichia papillosa*)

Marble screw-moss expresses sex only in Australasia.

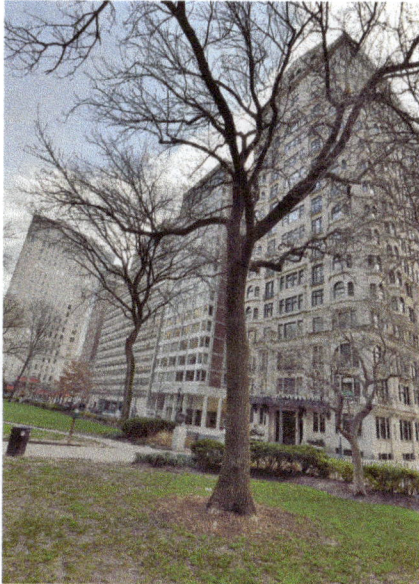

Amur cork tree (*Phellodendron amurense*), host to marble screw-moss.

Close-up of Amur cork tree, hosting marble screw-moss and another moss.

Marble screw-moss occurs in all continents.[1] Like dwarf bristle-moss, it is common on trunks of urban street trees.[2] Remarkably, it is known to make sexual structures and to reproduce sexually (i.e., to make spores) only in Australasia—Australia, New Zealand, and Tasmania. In Australasia, it makes male and female sex organs on separate plants.[3] Both sexes make specialized vegetative propagules in their leaves.[4] Raindrops disperse these propagules like those of umbrella liverwort. The propagules enable each sex to reproduce vegetatively on its own.[5]

A mystery is why marble screw-moss fails to express sex outside Australasia. I offer one hypothetical scenario. It presumes that marble screw-moss evolved expressing each sex on separate plants, as now in Australasia. It postulates that a lone spore dispersed far from its parental population and established a sexually isolated unisexual colony, perhaps on an island. This colony propagated exclusively vegetatively. Over time—possibly millions of years—sexual function in this population degenerated through mutations like those described on page 32 for threadstalk speedwell. After losing capacity for sexual expression, the colony spread globally while its sexual ancestral parent population endured in Australasia.

Translucent light green spherical propagules in leaves of marble screw-moss.

Sporadic Sex: silvergreen bryum moss (*Bryum argenteum*)

Silvergreen bryum moss propagates sexually or vegetatively depending on circumstances.

Silvergreen bryum moss is one of the most widely distributed plants on the planet. It populates cities, deserts, and mountains ranging from Brazil to Antarctica. In cities it grows on roofs, pavement, golf courses, sewage treatment plants, and old cars.[1]

Bryologists in Norfolk, England, hypothesized that small fragments of the moss adhere to shoes and tires, which disperse the moss. To test this idea, they created an artificial cinder pathway in the form of a cross, with a clump of moss placed in the center. They pressed a matchbook across the clump onto one axis of the cross every two days for three weeks. They found that the moss spread only along the axis of the cross in contact with the matchbook.[2]

Silvergreen bryum moss growing on concrete at the base of a brick rowhouse.

By contrast, bryologists attributed distribution of the moss in a valley in Antarctica to specialized vegetative propagules (bulbils) that float.[3]

Spore capsules of silvergreen bryum moss. The green capsule has yet to release its spores. The other two capsules have discharged their spores.

On putting greens, silvergreen bryum moss is regarded as a pest. In samples from putting greens across North America, it was found to consist of only one sex—female. Females of this moss are taller than males. On putting greens, mowers presumably fragment and disperse the tall females. In populations sampled as controls outside putting greens, this moss consisted of mixtures of male plants and female plants.[4] On a rocky outcropping in Brazil, some populations were all female while others were all male. Most were a mixture and reproduced sexually, making spores. Still others expressed no sex.[5]

In downtown Philadelphia, this moss thrives in diverse habitats. Colonies that reproduce sexually are rare, as evidenced by rarity of production of spore capsules. The dominant mode of reproduction appears to be vegetative. Perhaps its many modes of reproduction have helped silvergreen bryum moss spread across the city.

Chapter 1 // Nonflowering Plants

/ HERBACEOUS ANNUAL PLANTS /

Prostrate knotweed (*Polygonum aviculare*) rooted in a crack in a concrete walkway.

Mixed in with prostrate knotweed in the photo is a lighter green herbaceous annual, carpetweed (*Mollugo verticillata*). The minute size of flowers of both species is typical of self-pollinating herbaceous annuals.[1] Herbaceous annual plants like these can overwinter as seeds sheltered within cracks of concrete. They can grow in pavement cracks in part because they do not require space for bulky roots. Their growth is concentrated above ground and dedicated to production of seed. Other herbaceous annuals in the city predominantly cross-pollinate, as in the case of common ragweed. By contrast, jewelweed cross-pollinates or self-pollinates depending on local conditions. This chapter considers how mating systems of herbaceous annuals cope with urbanization.

Opportunistic Cross-Fertilization: common chickweed (*Stellaria media*)

Chickweed self-fertilizes by default, but it cross-fertilizes opportunistically.

Soon after a chickweed flower opens in the morning, only its male flower parts (anthers) are mature. It offers insects nectar and pollen. Later in the day its female parts (stigmas) mature, producing a flower that is functionally bisexual. Insects that have picked up pollen from anthers in the morning can now transfer this pollen to stigmas of bisexually mature chickweed flowers. If this transfer goes to a different chickweed plant, it results in cross-pollination.[1]

Common chickweed in a crack in a brick sidewalk.

If no pollinators arrive, chickweed has a backup that guarantees fertilization. Toward the end of the day, stamens bend inward so anthers deposit pollen on stigmas.[2] Self-pollination enables chickweed to make seed even when insect pollinators are absent. Like annual bluegrass, described in the introduction, a lone chickweed plant can sexually reproduce isolated in the heart of a city.

Chickweed bud and flower.

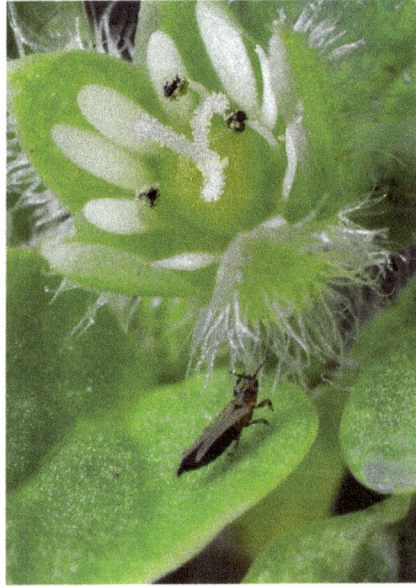

Pollinator (thrips).[9]

A single chickweed plant may produce up to 13,000 seeds, each 0.1 millimeter (0.004 inch) in length, a third to half of a milligram in weight, and viable 90 to 100 percent of the time.[3] For scale, 600 of these seeds would weigh less than a regular-strength tablet of aspirin (325 mg). Like annual bluegrass, chickweed seeds are dispersed by cars, shoes, and birds, and likely also by wind and water.[4] In Philadelphia, they lodge and germinate in pavement cracks. I suspect that these cracks facilitate cross-pollination in both species.

Darwin was puzzled that chickweed self-pollinates apparently without harm (from inbreeding).[5] Chickweed may avoid inbreeding depression because, like annual bluegrass, it is a hybrid species with four sets of chromosomes, two from each ancestral parent.[6] The extra chromosomes may protect chickweed from expression of harmful recessive genes.[7] In the event that inbreeding produces seeds with genetic deficiencies, chickweed's vast surplus of seeds can replace them. The result purges harmful recessive genes from the population.[8]

Chickweed flowers in different stages of development.

Fig. 1. Immature male and female structures.

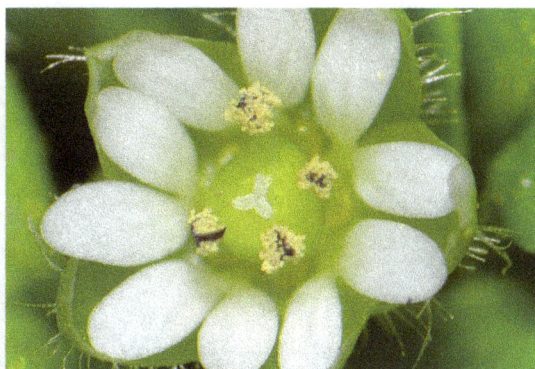

Fig. 2. Male structures (anthers) are all mature.

Fig. 3. Mature female structures (stigmas).

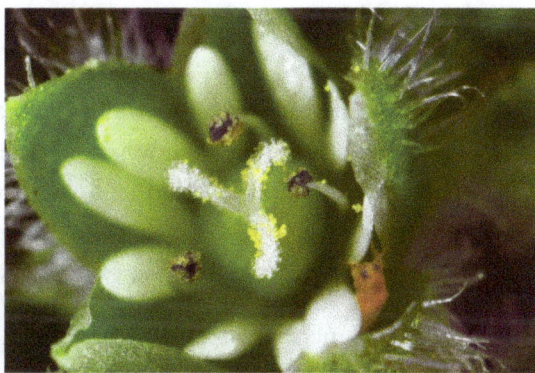

Fig. 4. Pollinated bisexual, with a thrips on lower right.

Fig 5. Bisexual self-fertilization.

The photos are of different flowers. The anthers (male structures) vary in number from three to five. They surround three white stigmas (female structures) that receive pollen in the center of the flower. Pollen is yellow.

In figure 1, the pink walls enclosing three of the anthers have yet to be shed; the three white stigmas are small and immature. In figure 2, the pink walls are all gone, and yellow pollen is visible on all four anthers. The flower is a functionally mature male; the three stigmas are still undeveloped. In figure 3, the stigmas are large and mature, but the anthers are missing. The flower is a functionally mature female. In figure 4, both anthers and stigmas are mature; a thrips has pollinated the flower, and its orange head is poking up from the lower right. In figure 5, stamens have bent inward; anthers are touching stigmas, and pollen is being transferred, effecting self-fertilization. Not shown are chickweed buds, which in cold weather may self-fertilize without opening.

Nectar Robbery: jewelweed (*Impatiens capensis*)

Nectar robbery in jewelweed bypasses pollination.

Jewelweed flower.

Bumblebee pollinating at the top of the flower.

Jewelweed produces ornate tubular flowers with nectar pooled deep in the bottom. Bumblebees struggle to claw their way down the tubes to access the nectar. Sometimes a bumblebee bypasses pollination. It bites a hole through the side of the flower, then accesses nectar directly through the hole rather than through the flower's natural opening at the top. It wins the reward of nectar but fails to do the job of transferring pollen.[1]

Other bumblebees may pollinate robbed flowers. In experimental studies based on simulated nectar robbery, Michael Zimmerman and Susan Cook at Oberlin College explored bumblebee pollination of jewelweed flowers that had been "robbed" compared to not robbed. They found that bumblebees visiting robbed flowers flew longer distances between flowers. Simulated robbery of flowers did not reduce their seed set. The researchers concluded that nectar robbery increases distance of transport of jewelweed pollen. It increases outbreeding, a primary objective of jewelweed's breeding system. They proposed that nectar robbery paradoxically rewards jewelweed with a reproductive benefit.[2]

In the bottom photo, the curve in the spur at the base of the jewelweed flower functions as a perch that supports the nectar robber. In one study, seed production and nectar robbing both increased with increased curvature of the spur.[3]

Nectar robbery through a hole bitten in the side of the flower.

Male-to-Female Sex Change: jewelweed (*Impatiens capensis*)

Jewelweed flowers switch from functionally male to functionally female.

Functionally male flower.

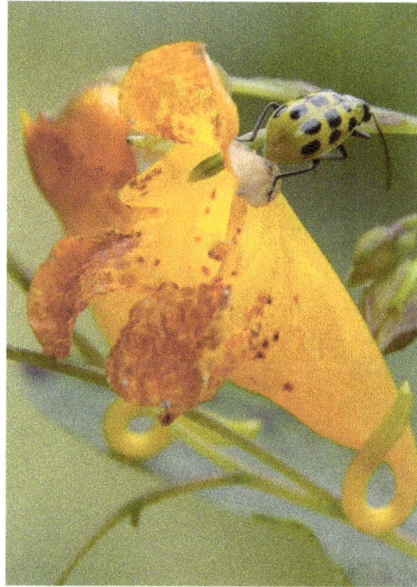

Functionally female flower.

A jewelweed flower is anatomically bisexual but functionally unisexual. When it first opens, it is functionally male, as in common chickweed. The white-colored anthers present pollen at the top of the opening of the flower, shown in the top left photo. When a bumblebee enters the flower, its hairy back rubs up against the anthers and picks up pollen, as shown in the bottom photo. Later, the anthers fall off, exposing female structures, such as stigmas hidden behind the anthers (top right photo).[1] When bumblebees laden with pollen from male flowers visit female flowers, they rub against stigmas, which receive their pollen. Unisexual function in this bisexual flower prevents a jewelweed flower from fertilizing itself. Molecular genetic markers have shown that most seeds from jewelweed flowers are products of outcrossing.[2]

Jewelweed illustrates "protandry": male function precedes female function.[3] Protandry occurs in common chickweed, cabbage white butterflies, and common garter snakes.

Bumblebee rubbing against pollen from functionally male flower.

Flowers That Never Open: jewelweed (*Impatiens capensis*)

Jewelweed produces flowers of two types; one type exclusively self-fertilizes.

Jewelweed in sunny wetland. These conditions promote profuse production of big, orange flowers. Not visible in the photo are small, dull flowers of a type that never opens and that only self-fertilizes.

This book illustrates jewelweed flowers as big, showy orange blossoms. Jewelweed produces a second type of flower, not shown. These flowers are small and unphotogenic. They stay closed, and they exclusively self-fertilize. An individual jewelweed plant produces flowers of both types at the same time.

The big advantage of the closed type of flower is that it requires neither pollinators nor mates to produce seed. Also, it is cheaper to produce, and it can make seed under relatively dry, shady conditions.[1] However, compared to the open type, the closed type produces self-fertilized progeny[2] of lower quality than the cross-fertilized progeny of open flowers. Seedlings from seeds of closed flowers are less competitive and less likely to survive the winter.[3] (Open flowers are functionally unisexual and cannot self-pollinate; however, a bee can accomplish the genetic equivalent of self-pollination by picking up pollen from a functional male flower. Once that flower or another flower on the same plant switches to being a functional female, the bee can pollinate the plant with its own pollen.)

Given the many tradeoffs between the two types of flowers, what determines which type jewelweed makes? Exposed to sunlight and moisture—ideal conditions for growth—jewelweed shifts its output in favor of the large, expensive flowers that open, as shown in the photo. Subjected to drought and shade—poor conditions for growth—it defaults to making the small, cheap flowers that stay closed, making inferior progeny.[4] I suspect that jewelweed's adaptability helps it endure in cities, where landscapes are prone to frequent disturbance affecting water and light.

Other plants that grow spontaneously in Philadelphia make flowers of two types, open and closed. They include: Venus's looking glass (*Triodanis perfoliata*), henbit (*Lamium amplexicaule*), common blue violet (*Viola sororia*), neckweed (*Veronica peregrina*), and prostrate knotweed (*Polygonum aviculare*, page 9).[5]

Quantitative Gender: common ragweed (*Ambrosia artemisiifolia*)

Sexual expression in common ragweed varies quantitatively.

Spikes of male flowers on common ragweed.

Ragweed plant with only female flowers.

Each of the many flowers of a common ragweed plant is either male or female. The ratio of male flowers to female flowers varies widely from one plant to another. A metric referred to as "gender" has quantified such variation,[1] but a more precise term is "sexual expression." In one study of sexual expression in ragweed, some plants made only female flowers while other plants made three times more male than female flowers.[2]

Female flower with two stigmas for receiving pollen.

What might cause sexual expression to vary? Controlled experiments regulating exposure to light showed that shade encouraged ragweed plants to produce flowers that are female.[3]

Male spike.

Why might shade favor female flowers? The female ragweed flower is a discrete, minute structure that makes just one seed, whereas the male ragweed flower is part of a tall spike consisting of many male flowers filled with pollen. Perhaps the solar energy required to make small female flowers is low compared to that required to make big male floral spikes.

In contrast to environmental determination of sexual expression in common ragweed, determination of sexual expression in umbrella liverwort (page 3) and white campion (page 20) is strictly genetic.[4]

Male-Male Competition: common ragweed (*Ambrosia artemisiifolia*)

Only one pollen grain can fertilize a female ragweed flower, but many begin the process.

Ragweed flowers. Male spike is on left. Female flowers are on right and below male.

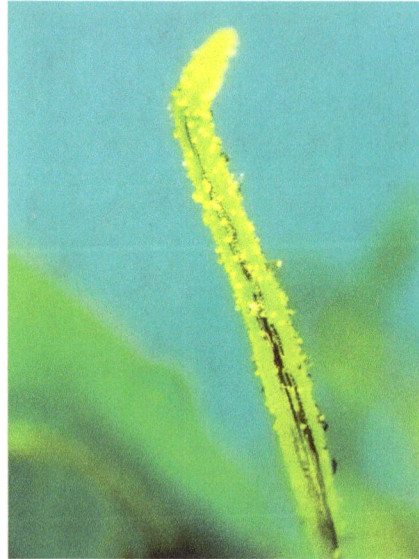

Ragweed pollen (yellow dots) coating stigma of female flower.

In the fall, wind-blown ragweed pollen commonly triggers hay fever. Ragweed is a mostly self-incompatible, outcrossing annual. It may produce more pollen than it actually needs for cross-pollination. Jannice Friedman and Spencer Barrett at the University of Toronto noted that a ragweed flower contains only one egg cell, so only one pollen grain can fertilize it; however, they found on average 30 pollen grains per stigma.[1] The stigma is the female flower part containing pollen receptors (photo on right). Because ragweed flowers synchronize their release of pollen grains, many pollen grains blown in from other plants onto a common ragweed stigma may arrive nearly simultaneously.[2] They must compete with each other to fertilize a lone egg cell. Chance may determine the winner.

Friedman and Barrett hypothesized that the winner in this reproductive lottery is the plant genotype that makes the most pollen, all else being equal.[3] According to this hypothesis, runaway competition may escalate production of pollen beyond quantities actually needed for pollination—until the self-reinforcing increases in pollen production generate disruptive costs, such as reduction in production of female flowers.

The astronomical potential of ragweed to make pollen was documented in the Midwest during the Dust Bowl years of the 1930s, when conditions for ragweed were ideal. In the air over Indianapolis, total ragweed pollen in one season averaged an estimated 7,263 grains per square centimeter, or 145 kilograms per square kilometer![4] Runaway competition among pollen grains may have contributed to this massive output of pollen. Alternatively, prodigious output of pollen may be an adaptation for long-distance pollination;[5] it may maximize cross-fertilization in populations where plants are widely scattered, as in cities. Both hypotheses may be correct.

Urban Sexual Isolation: common ragweed (*Ambrosia artemisiifolia*)

Urbanization can sexually isolate common ragweed.

Solitary common ragweed plant with no potential mates nearby. The species is mostly self-incompatible.

Amanda Gorton and colleagues at the University of Minnesota hypothesized that selective pressure in cities drives urban plants to evolve and adapt. They used urban ragweed to test this hypothesis. They collected urban and rural ragweed seeds and cultivated both under urban and rural conditions. They predicted that under urban conditions, fitness of ragweed grown from seed would be higher for urban compared to rural seed. A metric based on numbers of seeds and flowers quantified fitness. Contrary to prediction, urban and rural fitness of ragweed plants were both lower for plants grown from urban seed.[1]

In large populations, sexual reproduction typically generates genetic variability that fuels natural selection and evolution, boosting fitness. By contrast, in a small, isolated population, sexual reproduction may cause genetic drift and inbreeding, reducing fitness.[2] The city center core of Minneapolis where Gorton's team collected their urban seeds presumably confined common ragweed to small, isolated populations. Common ragweed has spread only recently in western compared to eastern North America, where it expanded after deforestation.[3] Recent colonization of common ragweed in Minneapolis may have contributed to low genetic diversity in that city.

The Minneapolis Code of Ordinances Relating to Health and Sanitation declares that weeds greater than 20 centimeters (8 inches) in height are "a nuisance condition and dangerous to the health, safety and good order of the city."[4] In cities, sexual isolation of plants like ragweed may be credited not only to buildings and pavement but also to people and policy.

Evolutionary Arms Race: field dodder (*Cuscuta campestris*)

Dodder is a flowering herbaceous vine that parasitizes other herbaceous plants.

Field dodder is a flowering parasitic vine native to North America. When mature, it has neither roots nor leaves, and it makes little chlorophyll.[1] Its stems are yellow or orange rather than green, and its flowers mostly white. It depends on its hosts for water, minerals, nutrients, and energy. It parasitizes a wide range of herbaceous plants.[2] In its search for a potential host, a dodder seedling senses volatile chemical cues and reflected light.[3] After contacting a prospective victim, it twines around its stems and taps into its circulation. It draws in water, nutrients, sugar, and large molecules. Dodder's germination root then dies, and dodder's host becomes its only connection to earth.[4]

Dodder produces short strands of RNA (microRNAs) that circulate into its host and suppress expression of host genes.[5] Dodder's RNA accomplishes this feat by targeting and selectively silencing host RNA (messenger RNA).[6] Dodder and its hosts both reproduce sexually, and both can make sexual genetic recombinants that provide weapons and counterweapons in evolutionary arms races.

For mugwort, dodder's arms races may exist more in theory than in practice. Mugwort is native to Eurasia (page 30). Its Eurasian ancestors may have had no coevolutionary contact with North American dodder, and thus no chance to evolve defensive weaponry. In Philadelphia, dodder may have blindsided mugwort. On the other hand, ancestors of mugwort may have had arms races with other members of the dodder genus, which is distributed worldwide.

Mugwort may be vulnerable for another reason. Its evolutionary potential may be low in its habitat along railroad tracks, shown in the upper photo. Here it spreads vegetatively. It may lack sexual activity required to compete in evolutionary arms races.

Field dodder climbing over mugwort.

Flowering vines of dodder twining around mugwort.

/ HERBACEOUS PERENNIAL PLANTS /

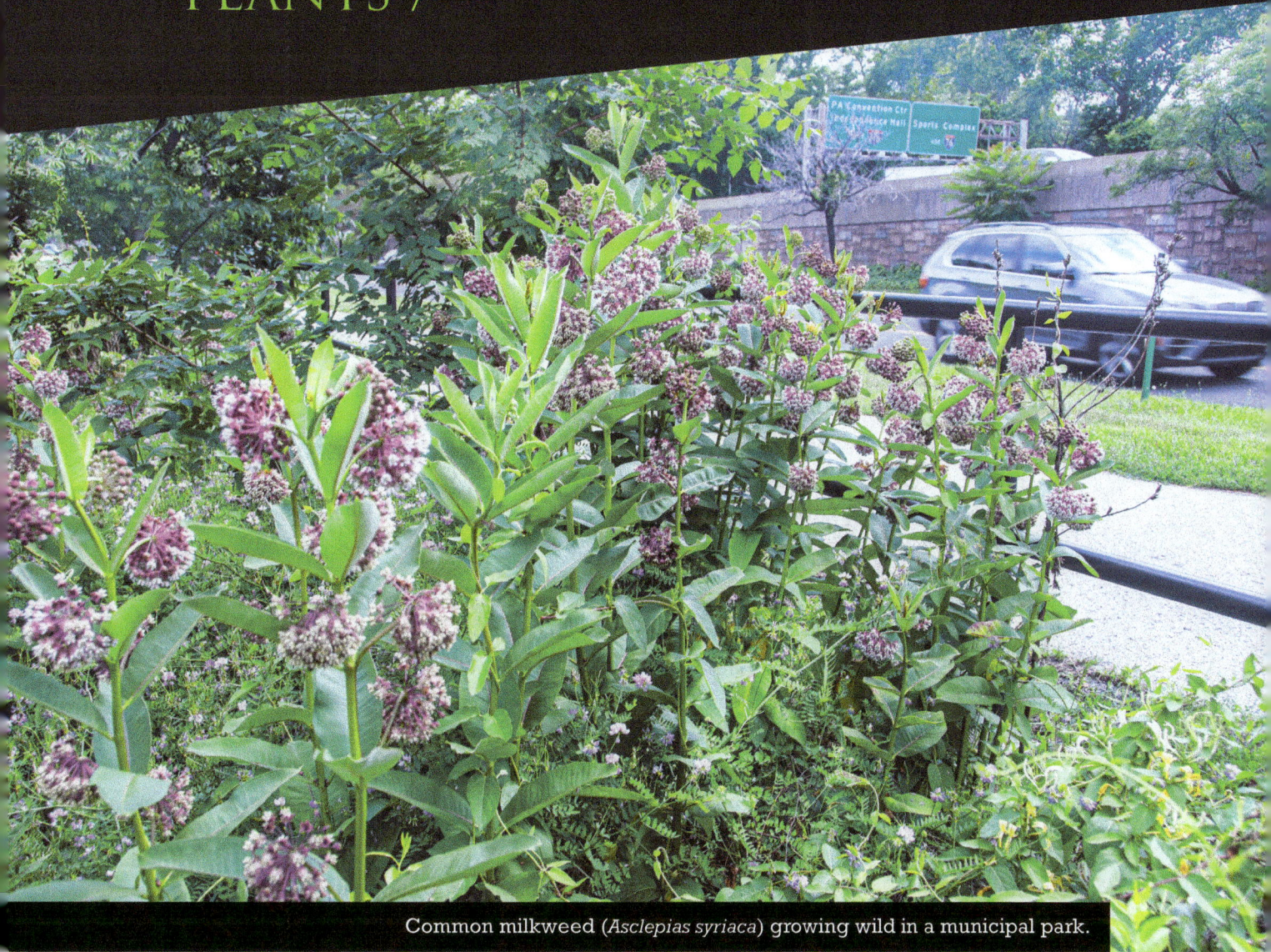

Common milkweed (*Asclepias syriaca*) growing wild in a municipal park.

Common milkweed is an herbaceous perennial with extensive horizontal roots that sprout shoots, producing vegetative clones. As shown in the photo, one or more of these clones can make big, showy floral displays that attract an abundance of pollinators. The mating system of milkweed blocks self-pollination and inbreeding within a clone.[1] Sexual isolation of a milkweed clone in the city might be expected to prevent a clone from making seed. This chapter explores mating systems of herbaceous perennials in the city. Some of these mating systems produce seed without fertilization; others produce no seed. In the case of herbaceous perennial plants, urbanization may support or undermine mating.

Sexually Transmitted Disease: white campion (*Silene latifolia*)

A sexually transmitted disease induces female white campion flowers to develop as male.

Female.

Male.

White campion is an annual or short-lived perennial introduced as an ornamental into North America from Europe.[1] It is one of a minority (5 percent) of flowering species with male and female flowers on different plants.[2] It hosts anther smut fungus (*Microbotryum lychnidis-dioicae*), a sexually transmitted parasite introduced from Europe.[3]

This fungus specializes in parasitizing white campion.[4] Insect pollinators carry its fungal spores from flower to flower.[5] After the spores germinate, the fungus disseminates throughout the plant's vascular tissue (xylem).[6] In female flowers, the fungus induces development of male sex organs (anthers) and suppresses development of ovaries; the result is anatomic sex reversal.[7] In male flowers, the fungus diverts production of pollen to production of fungal spores. Both male flowers and sex-reversed female flowers make fungal spores.[8]

The plant and its fungal parasite engage in an evolutionary arms race. At a molecular level, the parasitic infection activates or inactivates suites of genes whose expression in white campion can be detected by measurement of changes in levels of messenger RNA—a product of gene transcription. Some of these changes represent fungal manipulation of genes of white campion, while others represent white campion's defense. Su San Toh and colleagues at the Broad Institute of Harvard and the Massachusetts Institute of Technology documented these complexities in a paper published under the title, "*Pas de deux*: an intricate dance of anther smut and its host."[9]

In North America, white campion appears to be well equipped to engage its fungal foe. Genetic diversity of the fungus is low here compared to Europe.[10] While the fungus mostly self-fertilizes,[11] white campion exclusively cross-fertilizes (since its sexes are on separate plants). Cross-fertilization in white campion generates novel combinations of genes, potentially fortifying it with new weapons.

Hypotheses to explain ubiquity of sexual reproduction on earth cite evolutionary arms races against evolving enemies such as parasites.[12] The struggle between white campion and its sexually transmitted anther smut fungus is a case in point. Parasitism by field dodder, described in the previous chapter, is another.

Nocturnal Fertilization: white campion (*Silene latifolia*)

Moths pollinate white campion outside cities, but whether they do so within cities is unknown.

An urban habitat poorly illuminated at night.

Female, with fused calyx tube.

Nectaries in white campion flowers are deep inside a fused calyx tube 2.0 to 2.5 centimeters long—over four to five times the length of the tongue of the common eastern bumblebee, a visitor of white campion.[1] The neck of the calyx tube is too tight to admit bumblebees, so the nectar that bumblebees can access is only a fraction of that contained within the calyx tube. Other species of bumblebee, and also honeybees, visit white campion, but the length of even the longest tongues is less than half the depth of the calyx tube.[2]

White campion emits its scent exclusively or predominantly at night.[3] Its flowers open in the evening and close by midmorning.[4] Its pollen matures at night and reaches peak viability at midnight.[5] The solid white color of its flower viewed from above maximizes visibility at night. In studies of pollination in nonurban habitats, the dominant pollinators were moths.[6] Moths have tongues that are long enough to reach deep into the calyx.[7]

The common looper moth shown in the photo landed on an illuminated sheet in our garden. Attraction to the electric light diverted this moth from its normal activities, such as mating and pollination. In studies in Switzerland and England, artificial light suppressed pollination by nocturnal insects.[8] An open question is whether outdoor nocturnal lighting limits pollination and distribution of white campion in the city. Perhaps white campion in cities selectively populates dark refuges shielded from light pollution. Conceivably, its scent attracts nocturnal pollinators despite light pollution. Or pollination of white campion in the city might have shifted predominantly to the day, when diurnal pollinators such as bees visit it.

Common looper moth (*Autographa precationis*) attracted to an illuminated sheet in our backyard in downtown Philadelphia. It has been observed visiting white campion outside cities.[9]

Urban Self-Fertilization: creeping wood sorrel (*Oxalis corniculata*)

Flowers of creeping wood sorrel in Japan occur in two forms—urban and rural.

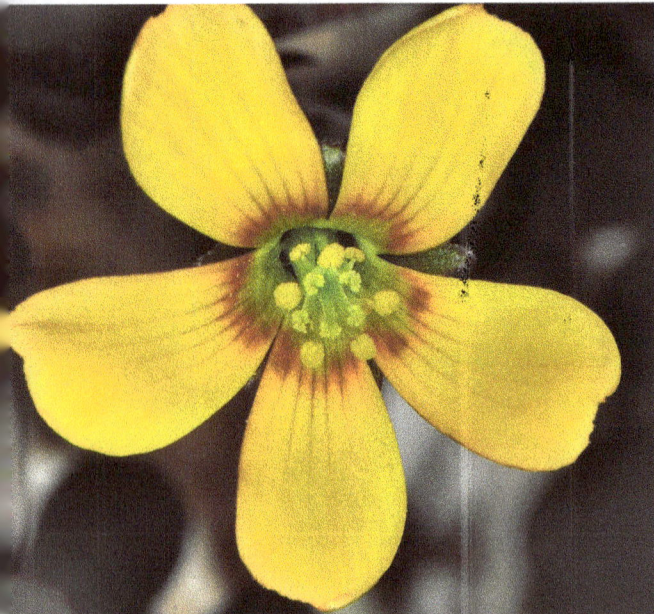

Creeping wood sorrel photographed in downtown Philadelphia. The yellow male (anthers) and pale green female (stigmas) parts in the center of the flower almost touch, facilitating self-fertilization.

If cities are prone to interfere with cross-pollination, they might be expected to favor self-pollination. Hiroyuki Shibaike, botanist at Kyoto University, and colleagues showed that creeping wood sorrel in Japan comes in two forms: urban and rural. In the urban form, male (anthers) and female (stigmas) parts of the flower touch or almost touch, favoring self-pollination. This urban form may also occur in disturbed habitats outside cities. By contrast, in the strictly rural form, these sexual parts are widely separated, favoring cross-pollination by insects. These sexual differences are inherited and are linked to a suite of other distinguishing traits.[1]

Self-fertilization in creeping wood sorrel allows the urban form to reproduce even when pollinators and prospective mates are absent. Self-fertilization also speeds up production of seed; it may advance seed set before mowing and weeding.

The plant is able to self-fertilize without evidence of inbreeding depression.[2] In Japan, creeping wood sorrel has eight sets of chromosomes. Like common chickweed, it may be protected from adverse effects of inbreeding.[3] Both urban and rural forms of creeping wood sorrel are self-compatible, meaning capable of self-fertilization.

Where the two forms diverged in the course of evolution is unknown. Creeping wood sorrel is a variable species cosmopolitan in distribution. In North America it grows as either an annual or a perennial. Recent evidence points to southeast Asia as its place of origin.[4] Cities may have favored the urban form regardless of where it evolved. The photo at the top of this page illustrates a flower of creeping wood sorrel in Philadelphia; the anatomy of the flower matches the urban form that Shibaike's team described in Japan.

Creeping wood sorrel on sidewalk in Philadelphia.

Seeds Without Fertilization: common dandelion (*Taraxacum officinale*)

Dandelion in North America makes seed without fertilization despite having bisexual flowers.

Dandelion in concrete parking lot.

Same dandelion as shown in the photo on the left.

Common dandelion has showy flowers that produce pollen and attract pollinators, but in North America it makes seed without pollination. Its offspring are maternal clones. Unfertilized germ cells develop into embryos, producing seeds.[1]

Common dandelion can adapt to diverse habitats in part because its clones are genetically diverse.[2] Most of dandelion's genetic diversity in North America has been imported in clones from Europe, where common dandelion is native.[3] In Europe, common dandelion makes seed with or without fertilization.[4] Pollen from dandelion clones can pollinate sexual dandelion (that makes seed with fertilization). Clone-sexual mating produces hybrids that are genetically diverse. These hybrids reproduce only by cloning.[5] Dandelion clones generated by this hybridization in Europe have carried their genetic diversity with them to North America.

Dandelion in turf.

In 1941, George Ledyard Stebbins, one of the leading botanists and evolutionary biologists of the twentieth century, concluded that an aging clone eventually becomes maladapted: it accumulates harmful mutations, and its clonality limits its power to adapt; it is doomed to extinction.[6] He did not know about "jumping genes" and other recently described sources of acquired but limited heritable diversity in dandelion clones.[7] But if Stebbins's prediction is correct, long-term survival of common dandelion in North America will require ongoing replacement of old clones by young clones introduced from Europe.[8]

Dandelion flower. It makes seed without pollination even though it attracts pollinators.

Gynogenesis: silver cinquefoil (*Potentilla argentea*)

Fertilization in most flowering plants is double. In silver cinquefoil it is single.

Silver cinquefoil in concrete parking lot.

Silvery underside of five-lobed leaf.

To make a seed, most flowering plants must be fertilized twice: a single pollen grain generates two sperm cells. One fertilizes an embryo-forming cell, and the other fertilizes an endosperm-forming cell. Endosperm will nutritionally support the embryo.[1] Silver cinquefoil is an exception. It fertilizes only endosperm-forming cells. Unfertilized embryo-forming cells produce offspring. Genes that silver cinquefoil passes on to future generations are exclusively maternal, despite reproductive dependence on males (for fertilization of endosperm).[2] This quasi-sexual form of reproduction is termed gynogenesis.

Some populations of silver cinquefoil in Europe reproduce by double fertilization,[3] but they have two sets of chromosomes (diploid) rather than the six sets of chromosomes (hexaploid) detected in North America.[4] With rare exceptions,[5] fertilization in hexaploid silver cinquefoil is single.

Pollination by a bee.

By forgoing double fertilization, silver cinquefoil sacrifices genetic recombination and evolutionary potential. What advantage might silver cinquefoil gain by sacrificing what most flowering plants go to great lengths to acquire? I suspect the answer lies in the genetic origin of this plant. Silver cinquefoil is a hybrid species; it is a product of hybridization of two ancestral parental species.[6] Cloning protects its successful hybrid genome from genetic recombination—it preserves hybrid vigor. At the same time, six sets of chromosomes theoretically protect silver cinquefoil from mutational harm.[7] Propagating in a pavement crack, a mother silver cinquefoil produces daughters who are just as well adapted as she to her pavement crack; she clones her strengths. On the other hand, exceptionally rare double fertilization may endow her progeny with at least some evolutionary potential—perhaps enough to help them adapt to cities.[8]

Chapter 3 // Herbaceous Perennial Plants

Mixed Modes of Mating: St. John's wort (*Hypericum perforatum*)

St. John's wort makes seed with self- and cross-fertilization, with single and double fertilization, and without fertilization.

A single plant of St. John's wort typically makes seed by more than one route. It may make seed by self-pollination as in chickweed; by cross-pollination as in jewelweed; without pollination as in North American dandelion; and by gynogenesis, as in silver cinquefoil. Fritz Matzk at the Leibniz Institute for Plant Genetics and Crop Research, Germany, and colleagues, identified 11 different pathways by which St. John's wort makes seed. But genetic control of these pathways eluded them.[1]

Why does St. John's wort use many systems of fertilization? One can guess: randomness of forces determining systems of fertilization may defy evolutionary pressure.[2] Diverse ecological conditions may favor diverse systems of fertilization. Complementary relationships among different mating systems may support multiple systems.

St. John's wort was introduced as an ornamental and medicinal plant into North America from Europe, where it is native.[3] In North America, the plant rapidly naturalized and spread. John Maron at the University of Montana and colleagues compared St. John's wort from populations in North America and Europe. They found that size and fecundity of the plant in North America adaptively evolved in response to differences in latitude.[4]

I suspect that St. John's wort's multiple methods for making seed help it endure in cities. Self-pollination enables the plant to make seed despite sexual isolation. Cross-pollination promotes genetic recombination and evolutionary potential. Cloning preserves adaptive genotypes. Each mode of reproduction has strengths that support the others. Together they are complementary.

St. John's wort.

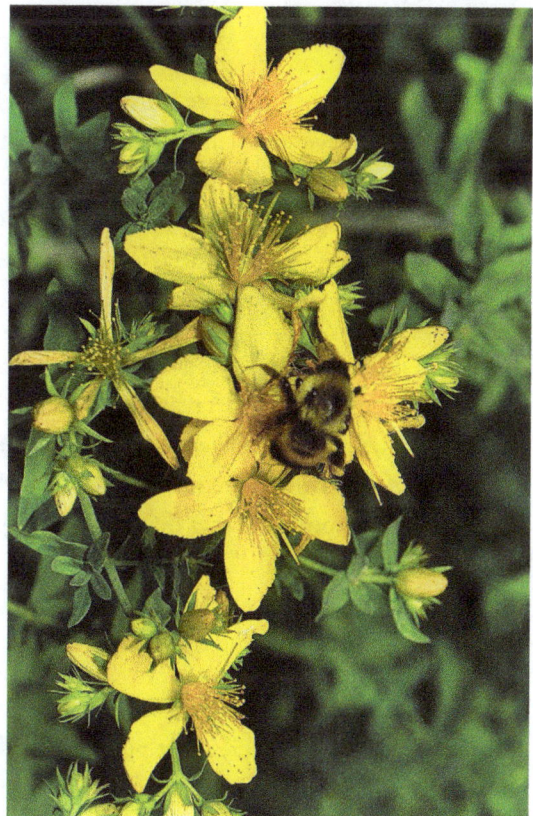

St. John's wort being pollinated by a bee.

Mating Types: yellow toadflax (*Linaria vulgaris*)

Yellow toadflax plants of the same mating type are genetically incompatible as mates.

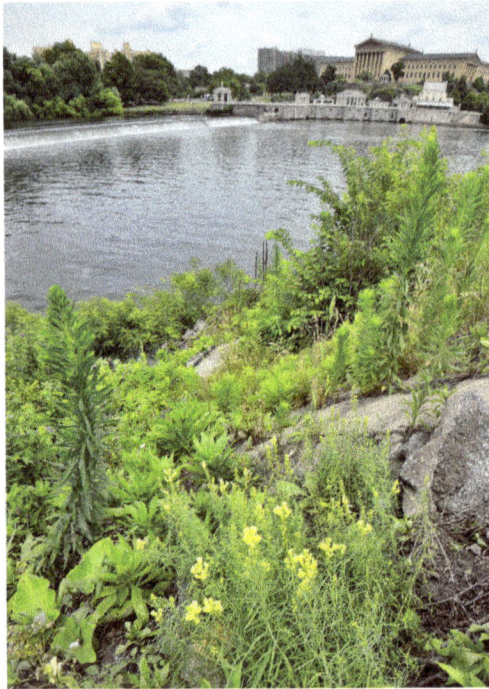

Wild colony of yellow toadflax.

Flowers of yellow toadflax.

Yellow toadflax spreads vegetatively from lateral roots that send up shoots.[1] These shoots form vegetative clones. Yellow toadflax has a mating system based on self-incompatible mating types. It permits pollination between two plants only if their mating types differ. This self-incompatibility system blocks pollination of plants that are genetically identical, such as those in a vegetative clone. It promotes outcrossing. Yellow toadflax has four identified mating types, which are determined genetically.[2] Bumblebees pollinate it.

Jacek Bartlewicz and colleagues in Leuven, Belgium, found that genetic diversity of clones of yellow toadflax decreased with urbanization. Seed production decreased in populations with lowest genetic diversity.[3] Ariana Longley in Ontario, Canada, found that hand pollination of urban yellow toadflax was less likely to produce seed when the pollen donor was from the same compared to a different urban population. Within an urban population, loss of genetic diversity likely reduced genetically compatible mates and pollination.[4] Urban declines in abundance of pollinators, and small, widely scattered floral displays, may also have decreased urban pollination and seed production.

A study of street plants growing spontaneously in metropolitan Paris found that the proportion of species that predominantly outcross declined with urbanization, as did the proportion that are self-incompatible and that are insect pollinated. Species with long, narrow tubular forms also declined with urbanization.[5] All of these traits characterize the mating system of yellow toadflax. Yellow toadflax may exemplify a mating system maladapted to dense urbanization.

Genetic Mixing: white clover (*Trifolium repens*)

White clover mates and genetically mixes in cities.

White clover flowering in a lawn.

White clover colonizing a gutter.

Unlike common ragweed and yellow toadflax, white clover carpets lawns in cities. It spreads by runners (stolons). White clover is a cosmopolitan perennial pollinated by insects. It is basically self-incompatible, requiring cross-pollination to make seed.[1]

Marc Johnson at the University of Toronto, and colleagues, found that genetic diversity of white clover did not decrease with urbanization. On the contrary, gene flow in the city was extensive. Rather than reducing fitness of white clover, urbanization drove adaptive evolution.[2]

Hans Verboven from Catholic University in Leuven, and colleagues, studied pollination in urban white clover. They found that pollinator visitation rates and seed set actually *increased* with increased urbanization. They hypothesized that either abundance of bumblebees increased with urbanization, or bumblebees concentrated in their study area.[3]

What might explain white clover's success overcoming urban isolation? White clover in Philadelphia offers clues. Here, white clover thrives in turf, which dominates green space downtown. Unlike ephemeral successional habitats, turf here is relatively stable from year to year. Groundskeepers tolerate white clover. It blooms below cutting blades of mowers.

Mowing grass increases exposure of white clover to sunlight and promotes its growth.[4] White clover growing in grass tolerates trampling. Shoes, tires, and birds disperse seeds of white clover.[5] Cities offer white clover rich opportunities to mix and mate.

Honeybee harvesting pollen from white clover.

Efficiency of Mating: common milkweed (*Asclepias syriaca*)

Packaging of milkweed pollen increases efficiency of pollen transfer.

Vertical slit between nectar cups.

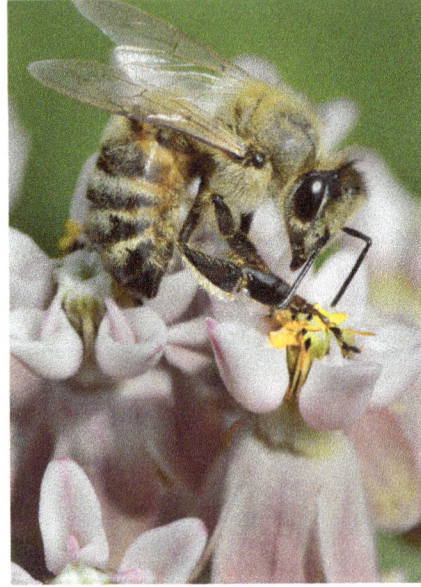

Insertion of an orange packet of pollen through slit.

Pollen is expensive to make. It contains not only nucleic acids (e.g., DNA), but also essential amino acids, fat, sugars, protein, minerals, and vitamins.[1] Common milkweed aggregates pollen and compresses it into distinctive flat packets (pollinia). The packets come in pairs linked by a sticky connector. When a bee sips nectar from a milkweed flower, a leg is prone to snag a connector and transport a pair of pollen packets. Stuck to a leg, the orange packets stand out. Later, as the bee flies to another milkweed flower, it accidentally inserts a packet into a slit that opens into the flower's stigmatic chamber. Slits are positioned vertically between nectar cups (left photo). The sticky connector now releases the packet, and transfer of pollen is complete.[2]

Pollen packets like these are unique to milkweed and orchid families.[3] Even though their transport on insects is cumbersome, they increase pollen transfer efficiency, reflected in the ratio of pollen grains to ovules. For outcrossing plants as a group, the ratio is 5,859 to 1; for common milkweed, the ratio is approximately 2 to 1.[4] This difference reflects more than increased efficiency of transfer of pollen. Bees, wasps, flies, beetles, and thrips eat pollen or feed it to their larvae,[5] but they cannot use milkweed pollen. Milkweed pollen packets stick to their mouthparts, entangle their legs, and do not fit into their pollen baskets.[6] In exchange for pollination, milkweed rewards insects only with nectar. (See photo of nectar theft on page 39.)

In cities, efficient transfer of pollen may help common milkweed pollinate; however, colonies of common milkweed take up space (as illustrated on page 19), and some people regard their appearance as weedy. Extirpation of common milkweed from urban neighborhoods can block cross-pollination despite the efficiency of milkweed's system for pollination.

Exotic Hybrid Vigor: common reed (*Phragmites australis*)

Crossing of Eurasian and native common reed created hybrids with aggressive traits inherited from both parents.

Common reed is a native perennial grass that grows to a height of 5 meters (16 feet). A Eurasian subspecies was accidentally introduced into North America in the nineteenth century in ballast dumps around port cities along the Atlantic coast. Ballast containing dirt and rocks from the holds of sailing ships was used as fill to convert marshlands into hubs for railroad and shipping. Assisted by railroads and highways, the Eurasian subspecies rapidly spread across North America. It was initially overlooked because it is practically indistinguishable from the native American subspecies, which grows naturally in coastal marshes (but is now comparatively rare).[1] In cities in the northeastern United States, common reed grows in dense stands in wetland and poorly drained vacant lots.[2]

In the western United States, common reed grows in disturbed wetlands and rivers around major urban centers. Spread has been credited to transportation and urbanization.[3] Hybrids of native and Eurasian subspecies of common reed were discovered along a waterway in Las Vegas, Nevada. Here, these hybrids have largely replaced the parental Eurasian subspecies. Each parental genotype passed on aggressive traits to its hybrid progeny. Hybrids threaten to infest wetlands along the length of the Colorado River.[4]

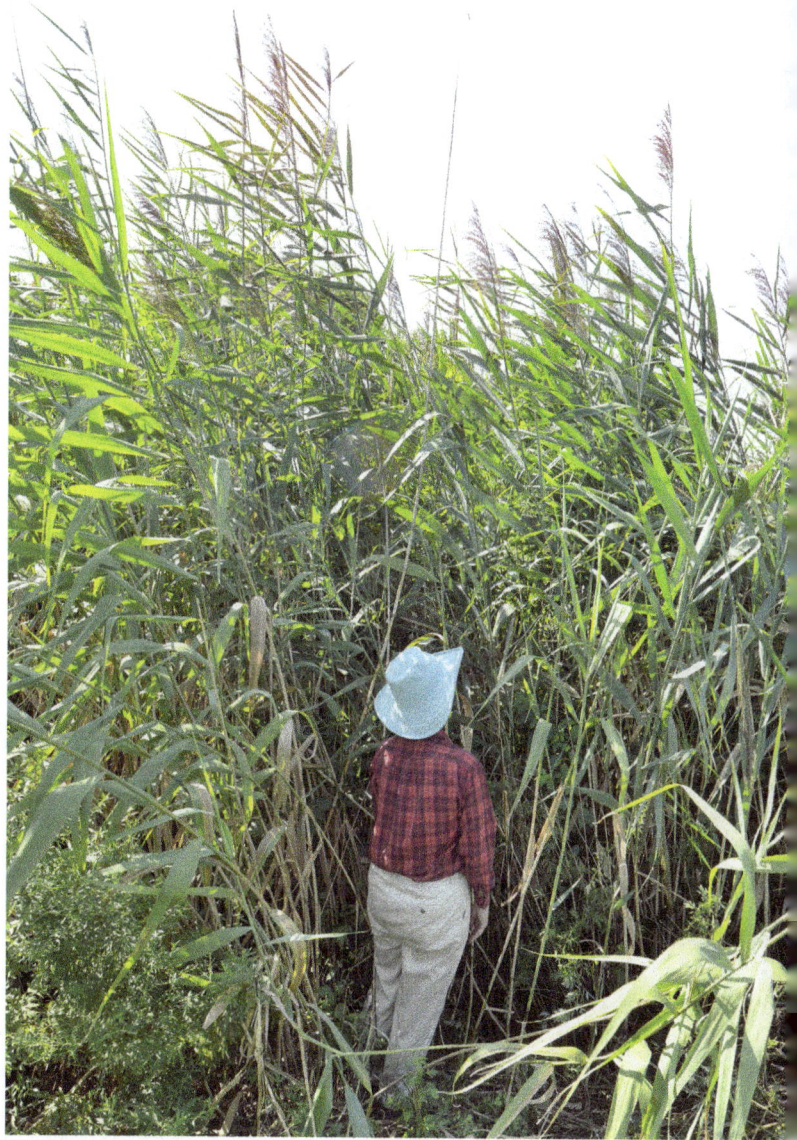

Common reed in a municipal park in Philadelphia.

Hybridization of common reed differs from that of mulberry (page 38). In the case of mulberry, both the exotic parent and its hybrids have endured, and the native parent has declined or disappeared. By contrast, hybrids of common reed have locally outcompeted the exotic invasive parent. Kristin Saltonstall of the Smithsonian Tropical Research Institute and colleagues presented this finding in a publication with the title, "What happens in Vegas, better stay in Vegas: *Phragmites australis* hybrids in the Las Vegas Wash."[5]

Variability in Fertility: mugwort (*Artemisia vulgaris*)

Fertility of mugwort differs widely in different regions.

Isolated colony.

Rhizomatous spread.

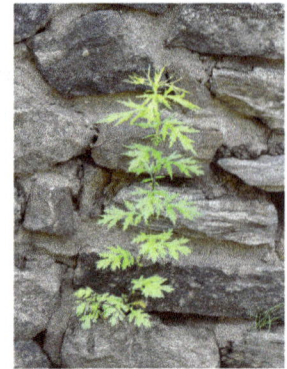

Seedling.

In *Wild Urban Plants of the Northeast: A Field Guide*, Peter del Tredici calls mugwort the "quintessential urban weed."[1] By contrast, in field and greenhouse studies during three years in Ithaca, New York, Leslie Weston and colleagues at Cornell found that mugwort produced no viable seed. Propagation was strictly vegetative, by rhizomes.[2] However, in downtown Philadelphia, mugwort spreads by both seed and rhizome. In studies in Europe, mugwort made on average more than 200,000 seeds per plant, and germination rates were 95 percent.[3] What might explain these wide differences in fertility?

In the nineteenth century, sailing ships imported mugwort mixed in ballast composed of dirt and rocks, which the ships dumped at port cities. Ballast dumps became beachheads for naturalization of plants introduced from around the world.[4] Mugwort evolved an aggressive variety that spread vegetatively by rhizomes.[5] It disseminated from cities through commerce and transportation networks.[6]

Mugwort is probably genetically self-incompatible.[7] Rhizomatous spread produces a clone of genetically identical plants. These plants are sterile unless pollinated by a genetically different plant. Outside of port cities, people spread mugwort mostly as vegetative clones.[8] Clonal spread from only a few sources could have established big monoclonal sterile populations, or superclones. Competitive exclusion of one clone by another could have also created superclones. Degeneration of sexual function in aging clones could have increased clonal sterility, as described in threadstalk speedwell (page 32). Plant toxins released by mugwort clones attack mugwort seeds and seedlings but spare mugwort rhizomatous shoots; these toxins effectively promote sterility.[9]

In Philadelphia, mugwort seedlings colonize cracks in pavement. Pavement blocks rhizomatous spread. In cityscapes, fragmentation of mugwort populations into small colonies surrounded by concrete opens up mugwort populations to genetic diversity. Repeated importation of mugwort from abroad likely introduced genetic diversity. I hypothesize that genetic diversity of mugwort in cities enabled cross-pollination and fertility. It overrode sterility within mugwort clones and helped mugwort transform into the "quintessential urban weed." This hypothesis is unproven. The genetics and breeding of mugwort in cities have yet to be studied.

Chapter 3 // Herbaceous Perennial Plants

Hybrid Speciation: salsifies (*Tragopogon* spp.)

Spontaneous hybridization of exotic salsifies in cities has created new species.

Yellow salsify (*Tragopogon dubius*).

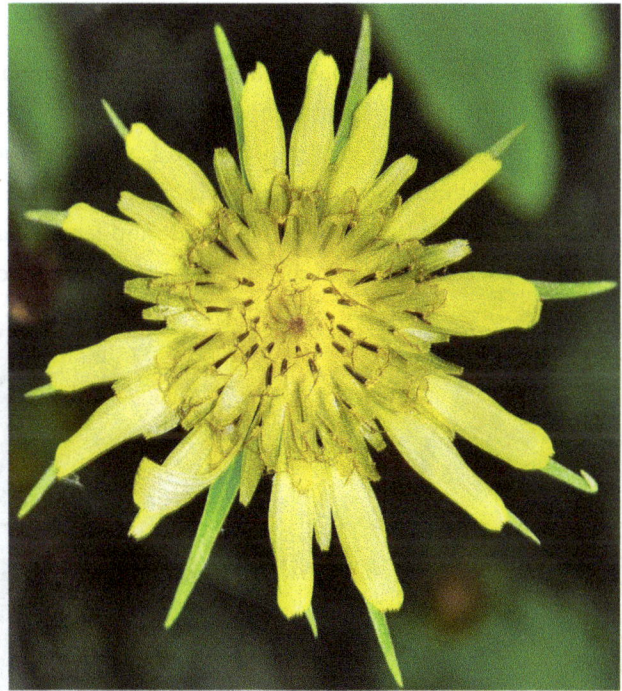

Yellow salsify.

In 1950, Marion Ownbey, botanist at the State College of Washington, described a new species of salsify (*Tragopogon*) in Moscow, Idaho. It bred spontaneously within the city, home of the University of Idaho. Two species of exotic weedy salsify (*T. dubius* and *T. pratensis*) in the city had naturally mixed and mated. Most of the hybrids were sterile, but at least one produced a hybrid that was fertile and bred true. This hybrid constituted a new species, which Ownbey named *Tragopogon miscellus*. Both parent species grow as biennials or short-lived perennials.

In the same paper that described this urban hybrid speciation, Ownbey reported yet another discovery: a second urban hybrid species, also the product of natural hybridization of two weedy exotic urban salsifies (*T. dubius* and *T. porrifolius*). He named this second species *Tragopogon mirus*. This hybrid species turned up in the city of Pullman, Washington, home of the State College of Washington, which employed Ownbey.[1] By 2009, hybrid speciation in salsifies had been documented in towns and cities in at least five western states.[2]

Each hybrid species has four sets of chromosomes, two inherited from each parent. Ordinarily, offspring of these salsifies inherit two sets of chromosomes, one set from each parent. An aberration in sexual reproduction (meiosis) increased the number of chromosomes inherited in the hybrid species.[3] In plants, an increase in chromosome number is a frequent pathway to speciation. Common chickweed, silver cinquefoil, and annual bluegrass exemplify hybrid species with extra sets of chromosomes.[4]

Lost Sex: threadstalk speedwell (*Veronica filiformis*)

Threadstalk speedwell reproduces without making seed even though its flowers are bisexual.

Threadstalk speedwell is a creeping perennial with attractive light-blue flowers. Endemic to the Pontic-Caucasian-Armenian mountains of Turkey and Georgia, it was introduced as an ornamental garden plant into England in the late eighteenth century. Escaping cultivation, it widely populated mowed lawns in North America, Europe, and Russia.[1] Its flowers are bisexual, containing both male and female parts, but the plant is self-incompatible: to make seed, it requires cross-pollination by a genetically compatible mate.[2] Outside its native range, such mates have been unavailable. In its introduced range in North America and Europe, the plant produces no viable seed.[3]

In regions where threadstalk speedwell has been introduced, the plant propagates exclusively vegetatively. It grows rootlets from creeping stems and from almost any shoots that touch the ground.[4] In cities and suburbs, lawn mowers scatter plant fragments, which then take root.[5] One might expect that after many years sexual structures that have ceased to function would degenerate.

Romain Scalone at the Swedish University of Agricultural Sciences and Dirk Albach at the University of Oldenburg confirmed this prediction. They showed that German

Threadstalk speedwell spontaneously populating a suburban lawn (see page x).

populations of introduced threadstalk speedwell produced fewer pollen grains than did flowers from native populations. The decrease in pollen grains correlated with deformities in male flower parts (stamens). They discovered one German population that makes no flowers. Scalone and Albach credited degeneration in sexual structure and function in threadstalk speedwell to natural accumulation of mutations.[6] Because the plants propagate strictly asexually, mutations harmful to sexual function do not harm the plants; on the contrary, mutations that block production of pollen might free up resources for structures that are more rewarding, such as rootlets.

Degeneration of sexual anatomy and function in threadstalk speedwell in its introduced range is likely to endure. Threadstalk speedwell illustrates the adage, "use it or lose it."

Chapter 3 // Herbaceous Perennial Plants

/ TREES AND WOODY VINES /

Three London plane trees (*Platanus x acerifolia*).

London plane trees are horticultural crosses between native and exotic sycamores. By the second half of the last century, 50,000 of these hybrid cultivars had been planted as municipal street trees in Philadelphia.[1] London plane trees have hybridized (backcrossed) with wild native sycamores there.[2] In California, hybridization with London plane trees has undermined the genetic integrity of a species of sycamore endemic to the region.[3] Hybridization of other exotic woody plants introduced into cities has undermined species that are closely related to them. This chapter examines actions and interactions of mating systems of urban woody plants.

Segregation of Sexes: box elder (*Acer negundo*)

Male and female box elder trees favor different habitats.

Box elder.

Native to North America, box elder is the most widely distributed maple on the continent. It ranges coast to coast across much of the United States and Canada.[1]

Box elder, like white campion, bears male flowers and female flowers on different plants. But box elder takes sexual segregation one step further: it segregates male and female trees. Although the sexes do spatially overlap, female box elder predominates along streams, while male box elder prefers upland.[2]

Theoretically, segregation of sexes into different habitats increases chances of reproductive failure. Box elder flowers are pollinated by wind. The likelihood that a pollen grain from a male box elder will pollinate a female box elder declines as the distance separating the two increases. At first glance, sexual segregation would appear to jeopardize pollination.

Segregation of sexes has an offsetting benefit: outcrossing. The farther plants are from their parents, the less likely they will mate with their brothers or sisters. Box elder pollen disperses four times farther under conditions that are dry compared to wet, so concentration of males away from streamside increases pollen dispersal.[3] Increased pollen dispersal offsets shortfalls due to sexual separation.

Seedlings of box elder.

Sexual segregation theoretically reduces sexual conflict. It insulates each sex from competition with the other, such as for water, sunlight, and space. It expands the geographic area and ecological habitat open to the species as a whole. It may generate evolutionary opportunity: compared to males, female box elders were found to produce leaves resistant to injury by insects and other herbivores.[4]

How does box elder achieve sexual segregation? The most plausible hypothesis is differential survival: better in dry areas for males and in wet areas for females.[5]

Chapter 4 // Trees and Woody Vines

Sex Change in a Tree: silver maple (*Acer saccharinum*)

Silver maple may switch sex from female to bisexual and back.

Male flowers.

Female flowers.

Silver maple is native to the eastern United States but widely planted as an ornamental.[1] It can be: (1) functionally female, producing only female flowers; (2) functionally male, producing only male flowers; or (3) functionally bisexual, producing male flowers and female flowers[2] (illustrated on page xii). It may switch sexual expression from one year to the next. A tree that is functionally female one year may the next year be functionally bisexual, and in subsequent years functionally female. In contrast, functionally male silver maple does not switch sex.[3]

Ann Sakai at Oakland University and Neal Oden at Stony Brook University reported that female silver maple trees of large size tended to switch to bisexual.[4] This switch could genetically benefit these big trees. For example, suppose a big female silver maple is especially well adapted. To propagate, it must receive pollen from another silver maple, which may be less well adapted. By switching from female to bisexual, the tree acquires the capacity to self-pollinate, increasing the chances that the genes its offspring inherit will be exclusively its own. Sakai and Oden also reported that big bisexual silver maples tended to clump, encouraging big trees to cross-pollinate each other.[5]

A switch from female to bisexual offers another potential advantage: it enables a female to propagate in the absence of a male. Propagation without a mate is advantageous in cities that promote sexual isolation. A switch in the reverse could be advantageous if a bisexual tree became maladapted due to environmental change. By switching from bisexual to female, a maladapted tree switches from mostly self-pollination to mandatory cross-pollination. The maladapted tree may now endow its progeny with outside genes.

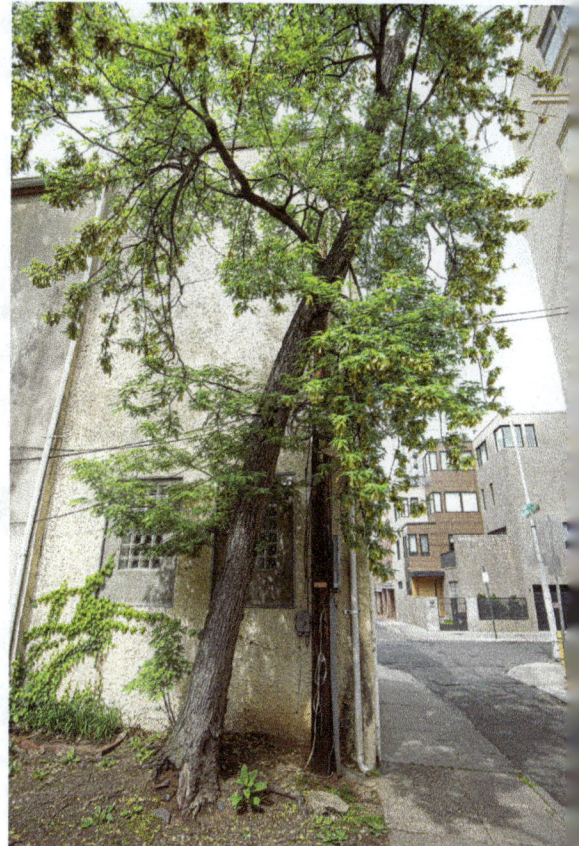
Female silver maple laden with winged seeds (samaras). The base of its trunk is rotting.

Sex Change Over Days: Norway maple (*Acer platanoides*)

In a single bloom season, Norway maple trees switch from producing male flowers to female flowers, or vice versa.

Norway maples.

Norway maple is widely planted as a municipal street tree. It has escaped cultivation and competes with native trees in the wild.[1] When a Norway maple starts to flower in the spring, it begins by producing either male flowers or female flowers. After a week to ten days these flowers wither, and it switches to making flowers of the opposite sex. At any time, it presents flowers of only one sex, preventing self-pollination. A tree that starts out as male can cross-pollinate a tree that starts out as female. A tree that has switched to male can cross-pollinate a tree that has switched to female.[2]

Studies of Norway maples have identified individual trees that have switched sexes twice in one season, or that have stayed the same sex. Exceptionally, both sexes have been reported to appear together on the same tree at the same time.[3]

The photos below show flowers from a Norway maple that started blooming with male flowers. During the five days between photos, it switched to making female flowers, and its male flowers withered.

Functionally male Norway maple flower, photographed April 9. The bright yellow structures on stalks are male (anthers) and contain pollen.

Functionally female Norway maple flower photographed on April 14 on the same tree that produced the male flower shown on the left. The yellow structures around the base of the flower are nonfunctioning male structures (anthers).

Exotic Male Interference: bittersweets (*Celastrus* spp.)

Male output of Oriental bittersweet (Celastrus orbiculatus) degrades female output of American bittersweet (Celastrus scandens).

Oriental bittersweet.

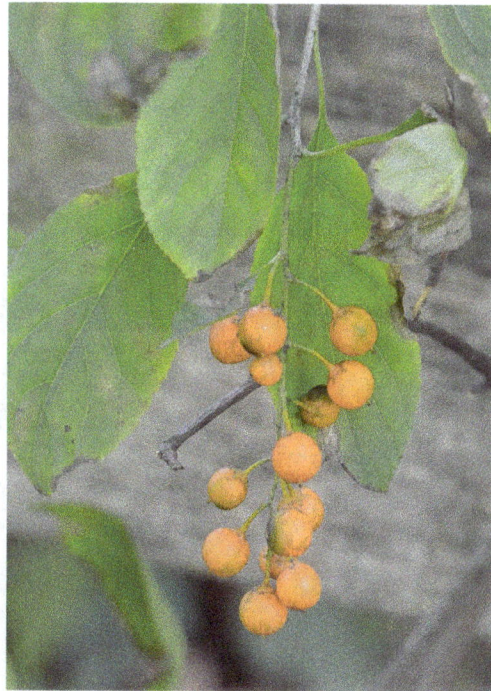

American bittersweet.

Oriental bittersweet is a woody vine introduced as an ornamental into North America from Japan in the second half of the nineteenth century. It produces attractive berries. Initially sold by a nursery in New York City, it escaped cultivation, naturalized, and spread throughout the New York City metropolitan area.[1] As the exotic bittersweet proliferated, native bittersweet (American bittersweet) declined or disappeared.[2] In Philadelphia, exotic bittersweet is widespread in naturalized parkland. I have found native bittersweet only under cultivation.

Pollen from Oriental bittersweet fertilizes American bittersweet. The direction of pollination is mostly in one direction: from exotic to native. The hybrid has low fitness: hybrid pollen is inviable, and hybrid seed set minuscule. Essentially, male output (pollen) from Oriental bittersweet diverts female investment (ovules) of American bittersweet to a reproductive dead end. Oriental bittersweet grows more vigorously than the native species. Spread of Oriental bittersweet amplifies its pollen output and its sexual disruption of American bittersweet.[3]

Decline in abundance of American bittersweet in the wild may be credited to a number of factors, including sexual interference by Oriental bittersweet.

Exotic Male Dominance: mulberries (*Morus* spp.)

Pollen from white mulberry (Morus alba) outcompetes pollen from red mulberry (Morus rubra) for pollination of red mulberry.

White mulberry (or a hybrid).

White mulberry, native to China, was introduced into North America starting in the seventeenth century to support domestic silk industries.[1] The tree escaped cultivation and widely naturalized, especially in cities.[2] Over the past century in the New York City metropolitan area, abundance of white mulberry has increased while abundance of red mulberry has declined.[3] (Despite the name, white mulberry and its hybrids bear fruit variable in color.)

Exotic white mulberry pollinates native red mulberry. As in the case of bittersweet, pollination is mostly in one direction: from exotic to native. In contrast to bittersweet, exotic mulberry and its hybrids are both more fit than the native species.[4] Exotic and hybrid pollen outcompete native pollen for pollination of native mulberry.[5] Male dominance of exotic mulberry and its hybrids threatens locally to extirpate red mulberry as a discrete genetic entity.[6]

Male flowers (catkins) of white mulberry explosively eject pollen at velocities over half the speed of sound. This motion is alleged to be the fastest reported for any living being. Ejection velocity of pollen from red mulberry has yet to be measured. Pollination is by wind.[7]

5

/ BUTTERFLIES AND MOTHS /

"Nectar theft" from common milkweed.

After sipping nectar, this Zabulon skipper (*Poanes zabulon*) flew off without picking up or depositing pollen. It "stole" nectar in the sense that it gave the milkweed nothing in return. The bagworm moth and the cecropia moth likewise offer flowers no rewards. These moths do not visit flowers. They do not eat. To mate and lay eggs they rely exclusively on energy stored from their larval stage. In cities where night-blooming flowers are scarce, such a strategy may be advantageous. On the other hand, refueling at flowers supplies the cabbage white butterfly with the energy it requires to disperse and repeatedly mate. This chapter explores mating strategies of urban butterflies and moths.

Aphrodisiacs and Anti-aphrodisiacs: cabbage white butterfly (*Pieris rapae*)

During courting, a male cabbage white applies aphrodisiac pheromone to his prospective mate. During mating, he instills anti-aphrodisiac pheromone into his mate.

Refusal posture of female cabbage white (left) rejecting advances of her suitor.

The cabbage white is one of the most abundant butterflies in North America. In the course of a year in most of its range, it has multiple broods—as many as eight in the southern United States.[1]

In the photo, a suitor is courting. He is showering her with aphrodisiac pheromone.[2] Her open wings and raised abdomen signal rejection (contrary to how human beings might interpret them). She has probably recently mated. During copulation, the male transfers his sperm in a package called a spermatophore. The spermatophore distends a receptacle inside her genital tract. This distension triggers a neural switch that turns off her sexual receptivity.[3] In addition, during copulation he instills into her an anti-aphrodisiac pheromone that temporarily renders her sexually unattractive.[4]

The female in the photo has attracted her suitor visually. He assessed her wing coloration from a distance. The coloration includes ultraviolet hues invisible to human beings. The particular pigments responsible for them signal her nutritional state and fecundity: the more light they reflect, the more attractive he finds her.[5] He identified her as female based on the double dots on her forewing.[6] She looked promising viewed from a distance, but his assessment changed once he had a chance to meet her close up. He fluttered beside her for only a couple of seconds before he flew off. If she mated again, it would be later. He would keep looking.

Males Emerge First (Protandry): cabbage white butterfly (*Pieris rapae*)

Protandry (emergence of males before females) enables female cabbage whites to mate while young.

Male, identifiable by his single dot.

Mating pair.

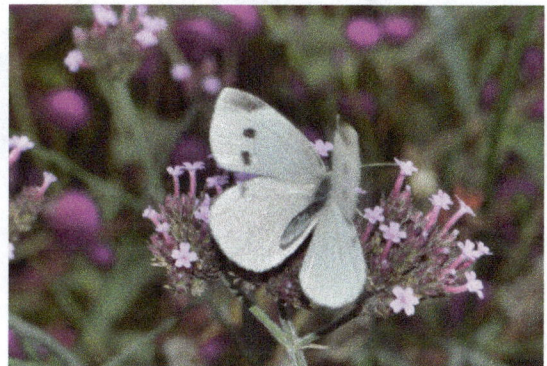

Female, identifiable by her double dots.

Because male cabbage whites emerge before females, males greatly outnumber an emerging female.[1] As soon as a female cabbage white hatches from her chrysalis, males attempt to mate with her.[2] Typically she mates close to her birthplace on her first day.[3] After mating she flies away, not yet mature enough to lay eggs.[4]

By securing her fertilization at the onset of her adult life, protandry frees her to disperse into unpopulated territory; she'll reproduce even if the new territory offers her no mate. In the course of their life span, cabbage whites in one study dispersed on average 2 kilometers from their place of birth.[5] Should she succeed in finding a mate in her new territory, sperm from her second mate, rather than her first, will likely fertilize her eggs.[6] Offspring from this second mate will inherit genetic mixing from a father who met their mother far from her place of birth. The mating system of the cabbage white may have contributed to the butterfly's success in populating big metropolitan areas.[7] By mating both before and after dispersal, urban populations of the cabbage white can achieve both high rates of fertilization and high genetic diversity.[8]

Average Is Best: male orange sulfur butterfly (*Colias eurytheme*)

Female orange sulfurs prefer male orange sulfurs whose size is midrange.

Male orange sulfur.

Theory predicts that females favor mates with traits signaling high quality. Female American toads mated preferentially with males that were bigger and sang longer and more often (page 89). Female house finches favored males that were brightest red (page 103). Male eastern gray squirrels in mating chases were bigger and dominant compared to other males (page 98).

Ronald Rutowski, lepidopterist at Arizona State University, studied mating success of orange sulfur butterflies. He evaluated success as a function of size, based on measurement of wings. He predicted that females would prefer larger males for several reasons. First, large size is a sign of genetic quality, since genes control the success of larval feeding. Second, large size may be advantageous for avoiding predation during copulation, when males in flight carry females. Finally, large size may translate into more nutrients transferred to females during copulation.

Contrary to prediction, Rutowski found that females preferentially mated with males that were midsized rather than exceptionally large. Similarly, males preferred midsized rather than exceptionally large females. He explored possible explanations for why orange sulfurs might prefer mates that are midsized, but he found all to be inadequate.[1]

Why exceptionally big wings depress sexual attractiveness in the orange sulfur is a mystery. One hypothesis is that the butterfly's flight control and other systems are optimized to wings that are midsized rather than exceptionally large. By contrast, strikingly small body size affords certain fish a sexual advantage, as described on page 93.

Chapter 5 // Butterflies and Moths

Male Territoriality: red admiral butterfly (*Vanessa atalanta*)

The red admiral defends territory based on urban landmarks.

Territorial red admiral displaying on a brick sidewalk downtown.

Red admirals in Philadelphia are one of the few species of butterfly that lands on human-made objects. Royce Bitzer and Kenneth Shaw, insect ecologists at Iowa State University, studied the behavior of red admirals on campus. They found that starting around 5 p.m. male red admirals began defending territories. Each territory included one or two fixed spots where a male spent most of his time. These spots were usually on the ground such as a sidewalk, occasionally on the walls of buildings, or less often on a bush or low branch of a tree. Territories were along sunlit, west-facing walls; sunlit lawns; or edges of walkways. They measured up to 13 meters in width and up to 24 meters in length.[1]

A resting male red admiral would leave his spot, patrol his territory, and return to the same spot. He would fly up underneath intruders or rush at them or hover above them. He would chase them in close pursuit out of his territory. He would also attack objects, both animate and inanimate, that entered his territory. These included birds or other insects, or even falling leaves. He would dart at them aggressively and repeatedly.

Shaw and Bitzer never observed red admirals in their territories mating, feeding, or laying eggs. They identified the sex of territorial red admirals by examining genitalia of captives. All were male.

In contrast to dragonflies (page 58), the male red admiral is not a predator and does not have mandibles or raptorial legs. Tsuyoshi Takeuchi at Osaka Prefecture University hypothesized that apparent territorial contests involving male butterflies of the same species are actually not adversarial, but rather exploratory. According to his "erroneous courtship hypothesis," apparent contests involving males are attempts by each to establish the sexual identity of the other.[2] This hypothesis would be difficult to test in the case of the red admiral in the wild. Human beings cannot distinguish the sexes of red admirals in flight, and mating occurs outside of territories defended.

Red admiral perched on a wrought-iron railing.

Sex-Limited Mimicry: eastern tiger swallowtail (*Papilio glaucus*)

The female eastern tiger swallowtail butterfly comes in two forms, one of which is mimetic.

Pipevine (*Aristolochia macrophylla*) cultivated in a city garden. Various species of pipevine are host plants of larvae of the pipevine swallowtail, a poisonous butterfly that black female tiger swallowtails mimic.

Male tiger swallowtail butterflies have black and yellow stripes. Females of this species come in two forms (shown on the next page): a tiger-striped form, which resembles the male; and a black form, which resembles a poisonous butterfly, the pipevine swallowtail (*Battus philenor*). The resemblance protects the black form from attack by birds that have learned to avoid pipevine swallowtails. The black form is a mimic, and the pipevine swallowtail is its model.[1]

Why is this mimicry limited to females? Three forces may be at play. (1) *Female preference*: Experiments in which male eastern tiger swallowtails were painted black (like mimics) or tiger-striped showed that virgin females preferred males painted tiger-striped. This female preference was independent of female color form.[2] (2) *Predation by birds*: Birds that prey on swallowtails selectively attack females, which are vulnerable to predation when they lay eggs. Males lack this particular vulnerability and so lack most of the benefit from mimetic protection.[3] (3) *Sex-linked genes*: Genes determining the black form in the female eastern tiger swallowtail reside on the female sex chromosome. Daughters are nearly always the same color as their mothers. Females have only one copy of this sex chromosome, which they pass on exclusively to their daughters.[4]

The prevalence of the black female form relative to the tiger-striped female form increases in geographic regions where the pipevine swallowtail and its host plant (pipevines, *Aristolochia* spp.) are most abundant.[5] Paradoxically, the black form of the female eastern tiger swallowtail is common in Philadelphia, where the poisonous model is rare. Perhaps genes for the black form are fixed in these urban populations, or the black form immigrates into the city. Whatever may be the explanation, the persistence of the black form in cities where models are rare makes a point: evolutionary forces outside the city can determine sexual expression within the city.

Tiger-striped form of the female tiger swallowtail.

Black form of the female tiger swallowtail.

Sequestered Females: evergreen bagworm moth (*Thyridopteryx ephemeraeformis*)

The female evergreen bagworm mates inside her bag.

Adult male bagworm moth.

Adult egg-laden female, outside her bag.

The male evergreen bagworm conceals his abdomen in a portable bag, a silken sack covered with bits of his host plant. In late summer, he pupates in his bag. Two to four weeks later he emerges as a moth. Guided by sex pheromones detected by his feathery antennae, he flies to a female inside her bag. He will not eat, and he will live as a moth for only 24 to 48 hours. He lands on her bag and finds an opening.[1]

Bagworm bag.

While he metamorphoses into a moth, she metamorphoses into an adult stage resembling a grub, with no wings, antennae, or ornamentation. Her legs and mouthparts are vestigial and her eyes minute. When ready to mate, she splits open her pupal case, but she stays inside her bag.[2] From her thorax, she releases sex pheromone-scented hairs, which she expels through cracks in her pupal shell toward a hidden opening in the bottom of her bag.[3] Her mate passes the tip of his abdomen through this opening and telescopes his abdomen inside her bag until his genitalia meet hers. They copulate inside her bag while his head, thorax, and legs are outside.[4]

After mating, she discharges 600 or more eggs inside her bag and shrivels up; her eggs had constituted up to 90 percent of her mass.[5] Now she works her way out of the bag, seals the opening, drops to the ground, and dies.[6] In spring, her eggs will hatch into caterpillars that leave the bag. They will soon make new bags and develop as bagworms. Sequestration of adult females in their bags will protect them from urban dangers that kill other nocturnal insects. These dangers include electric lamps, vehicular traffic, birds, and bats.

Chapter 5 // Butterflies and Moths

Long-Distance Sex Pheromone: cecropia moth (*Hyalophora cecropia*)

Male cecropia moths have tracked plumes of sex pheromone across urban landscapes at night.

The cecropia moth is a giant silk moth native to North America. It thrives in urban residential areas.[1] Gilbert Waldbauer and James Sternburg, entomologists at the University of Illinois at Urbana, studied its mating behavior in metropolitan Champaign-Urbana. They found that most eggs from mated siblings failed to hatch, while over 90 percent of eggs that were outbred hatched. Yet female cecropia moths laid their eggs close together in large batches, potentially producing siblings that inbred.[2]

Male cecropia moth.

They later showed that male and female moths hatch from their cocoons in the morning but defer mating. At dusk, males disperse while females stay, waiting several hours before they release their sex pheromone. The moths do not mate until just before dawn the following morning—allowing plenty of time for males to scatter. The investigators marked and released over a thousand males and recaptured slightly more than a third in traps baited with virgin females. Some of the males likely flew to wild females not in cages. Twenty males flew into traps nearly 7 kilometers or more from their points of release.[3]

By emitting their sex pheromone close together in time and space before they disperse,[4] female cecropia moths collectively reinforce each other's sexual signals. They maximize the distance over which they can attract males. The big feathery antennae of males also help to maximize this distance. With broad, powerful wings capable of flight over long distances, the male cecropia moth is well equipped to overcome urban sexual isolation.

Feathery antenna of male cecropia moth.

Infertile Mating: ailanthus webworm moth (*Atteva aurea*)

Infertile mating induces female ailanthus webworm moths to mate again.

Ailanthus webworm moth feeding on late boneset (*Eupatorium serotinum*).

Ailanthus webworm moth attracted at night to an illuminated wall facing our garden.

The ailanthus webworm moth feeds and mates during the day. Its showy colors alert potential predators of its unpalatability. Its conspicuousness may also assist in courtship. Its larvae feed on tree of heaven (*Ailanthus altissima*), an abundant urban tree and a likely chemical source of its unpalatability. The moth begins laying eggs on this plant just before dusk and continues into the night. It flies to electric light.

The female ailanthus webworm moth commonly mates more than once. Orley Taylor, insect ecologist and reproductive biologist at the University of Connecticut, questioned what induces females to mate two or more times. He suspected that a female webworm moth is more likely to mate a second time when sperm from her first mating is insufficient for fertilization. To explore this hypothesis, he observed nearly 700 copulations of webworm moths. In some of his experiments, he assessed fertility by monitoring egg laying and egg viability. In others, he dissected mated females and examined their reproductive systems. His findings demonstrated that sperm deficient in quantity or quality induced females to mate again. He suggested several possible humoral and neural pathways that might inform a female of the adequacy of the sperm.[1]

/ BEES, WASPS, AND ANTS /

Eastern yellowjacket (*Vespula maculifrons*) queen.

The black dots on the abdomen of an eastern yellowjacket queen distinguish her from males and workers (females that are not queens). The photo was taken in mid-April. This queen recently emerged from hibernation and is looking for a nesting site, such as a hole in the ground. She mated in the fall. If she succeeds in nesting, she will lay eggs and raise a brood, which will produce workers. In late summer and fall, her brood will produce sons and future queens. Her life cycle is similar to that of other social bees and wasps included in this chapter. While colonies of social insects in the city appear orderly, sexual conflict within them can disrupt the peace. This chapter examines some of these conflicts.

Matricide: bald-faced hornet (*Dolichovespula maculata*)

Bald-faced hornets sometimes kill their queen.

Nest of bald-faced hornets outside a rowhouse downtown.

Face of bald-faced hornet.

Hornets, yellow jackets, and bumblebees sometimes kill their queens. Workers in nests of these stinging insects are female and potentially able to lay viable eggs. Workers differ from queens in that they are all virgins; any eggs that workers lay are unfertilized. But these unfertilized eggs can produce sons. Workers that kill their queen can take over reproduction of the colony and pass on their genes through sons.[1]

Workers are all daughters of the queen (unless the current queen has replaced an earlier queen). If the queen has mated only once, all workers are sisters; these sisters are more closely related to one another than to their queen. The sisters share a reproductive conflict of interest with the queen.[2] If the queen has had more than one mate, some of the workers will be only half-sisters. Based on lower genetic relatedness of workers, the risk of matricide is expected to be lower under a queen that has had multiple mates (polyandry) rather than only one (monogamy). Studies comparing matricide in colonies with queens who have had multiple mates versus only one mate have borne out this prediction.[3]

All workers in a nest of bald-faced hornets have the same paternity. Roger Akre and Elizabeth Myhre at Washington State University described the killing of a queen in a bald-faced hornet's nest that they had transplanted into a viewing box with a glass bottom. They noted that workers had started to lay their own eggs in the nest despite the presence of the queen. The fateful episode began as follows: Contrary to her usual behavior, the queen wandered out of her nest, but she remained within the nest box. A lone worker grabbed her and tried to sting her. She bit and tried to sting the worker. A second worker joined the fray. The commotion aroused over three dozen excited workers who ganged up on the queen. The queen was stung to death. Akre and Myhre believed the excited worker and not the queen attracted the fatal mob. An unanswered question is whether reproductive rivalry ignited the initial conflict between worker and queen.[4]

Chapter 6 // Bees, Wasps, and Ants

Polyandry: common eastern bumblebee (*Bombus impatiens*)

Queens of the common eastern bumblebee have one or more mates.

In a paternity study of colonies of the common eastern bumblebee, two of ten queens were found to have mated with more than one male.[1] By contrast, queens of most species of bumblebee are strictly monogamous; mates of these monogamous queens inseminate them with a mating plug containing not only sperm but also a compound (linoleic acid) that prevents the queen from remating.[2]

Polyandry may endow queens of the common eastern bumblebee with benefits beyond suppression of worker rebellion. In experiments, artificial insemination was used to simulate polyandry and monogamy in bumblebee queens (of a European species). Compared to monogamous queens, polyandrous queens produced colonies with lower rates of parasitic (trypanosome) infection.[3]

In contrast to species of bumblebee that have declined in the United States, those that are stable or increasing were found to have higher genetic diversity and lower rates of a fungal disease. The common eastern bumblebee is in the group with higher genetic diversity.[4] Polyandry presumably contributes to this genetic diversity.[5]

Documentation of mating in the common eastern bumblebee is oddly rare, given the species' abundance. The photo shows a mating aggregation that Mary Anne Borge photographed in her garden in Lambertville, New Jersey, on a cool day in November. She watched the cluster of bees for an hour as males came and went. Her series of high-resolution close-up photos shows at least one male in copulation with the queen.[6]

Low temperature can prevent bumblebees from flying. This species of bumblebee stops foraging at ambient temperatures below 16 degrees Celsius (61 degrees Fahrenheit).[7] Perhaps low temperatures grounded the queen but not her suitors, whom her pheromone attracted.

Among the bumblebees in the photo, only the queen will overwinter. She will select a sheltered spot for her hibernaculum, probably a hole in the ground. In spring, she will emerge to establish a colony. If she has mated with more than one male, her colony will inherit genetic diversity and perhaps also resistance to disease. Both these traits might help her colony cope with urbanization.

Photo by Mary Anne Borge

Four males on a queen of the common eastern bumblebee during a cool day in November. One male is copulating.

Female Usurpers: common eastern bumblebee (*Bombus impatiens*)

Female usurpers attempt to kill queens of the common eastern bumblebee.

Common eastern bumblebee worker with pollen baskets filled with pollen.

Matricide has not been reported in the common eastern bumblebee. But outsiders may invade a hive and engage the queen in mortal combat. They are rival queens of the same species, or parasitic queens of other species. Both are potentially lethal usurpers who violently attempt to seize control of the hive. Both are black and yellow like her. If successful, each will kill her unless she escapes, and the hive will continue as a matriarchy with the usurper in control.[1]

If the usurper is a rival queen of the same species, workers (daughters of the deposed queen) will raise offspring from eggs laid by the usurper. If the usurper is a social parasite such as the lemon cuckoo bumblebee (*Bombus citrinus*), the usurper will kill many of the workers and enslave those that remain in the hive. The workers will be enslaved in the sense that they will raise offspring of a mother who is neither their own mother nor their own species. The lemon cuckoo bumblebee may allow enslaved workers to raise sons of their deposed queen.[2] By allowing production of these sons, the lemon cuckoo bumblebee indirectly supports her own progeny, who ultimately depend on propagation of the common eastern bumblebee. Unlike daughters, sons cannot sting; they cannot kill the usurper or her sons or daughters.

The usurper subdues workers of the deposed queen by "mauling" them: she grabs them as if to sting them but stops short of actually inserting her stinger.[3] Mauling by the lemon cuckoo bumblebee suppresses maturation of worker bumblebees' ovaries.[4] A contact pheromone (dodecyl acetate) that protects cuckoo bumblebees from attack by workers has been identified in another species of cuckoo bumblebee.[5]

The lemon cuckoo bumblebee controlling a hive has no baskets for harvesting pollen. She produces no workers of her own. She depends on workers of the deposed queen for foraging, raising her young, and defending the hive. She populates cities because the common eastern bumblebee, one of her host species,[6] thrives in them.

Male Stuffing: European paper wasp (*Polistes dominula*)

Female European paper wasps stuff male European paper wasps headfirst into empty cells of their nests.

Female.

Male.

Workers of the European paper wasp stuff males headfirst into empty cells of their nests by grappling, biting, and threatening to sting. Later, they restuff them; they bite and push the abdomens of males that they have already stuffed. Restuffed males in one study stayed stuffed on average over one hour. By male stuffing, workers temporarily deny males a chance to poach food the workers have brought in for the nest's larvae. When not stuffed into a cell, a male can forage on his own outside the nest; but he may prefer to eat in.[1]

Male European paper wasps stuffed headfirst into empty cells of a nest. Their legs and abdomens stick out. All the wasps that are not stuffed are females. The female in the top center is the queen (or "dominant female"); the other females do the stuffing.

Leks: European paper wasp (*Polistes dominula*)

Male European paper wasps congregate in leks—assemblages dedicated to attracting females.

Male European paper wasp feeding on nectar. He resembles a yellowjacket.

During summer and fall, male and female European paper wasps seeking mates disperse from their natal nests. How do opposite sexes find each other?

Compared to other insects, the male European paper wasp lacks adaptations for finding mates. He cannot sing like a cicada or flash like a firefly. He does not have big, colorful wings like those of butterflies. He lacks brilliant iridescent cuticle like that of metallic green sweat bees. Unlike the cecropia moth, he does not have fine feathery antennae for detection of sex pheromone over long distances. He does not go to nests of emerging virgin females, as do male cicada killer wasps; nor does he meet his mate on a specific host plant, as does the locust borer beetle. His black and yellow stripes are inefficient for visual identification. He is part of a large mimicry complex with many similar-looking species, including yellowjackets, cicada killers, and flower flies (syrphids).

In some animals, males collectively advertise their presence. They cluster in a site where they wait and put themselves on display for local females. The sexual display from the group as a whole exceeds that of any member alone. After attracting females, males vie for mates, and females pick and choose. Theoretically, such mating behavior brings the sexes together, avoids inbreeding, minimizes travel, and selects vigorous males. This male aggregation is a lek.[1]

Male European paper wasps form leks on conspicuous sunlit landmarks such as metal utility poles or tree limbs silhouetted against the sky. Males compete for perches high on these landmarks, which are located away from nests. Within the lek they defend their perches and surrounding territory from other males.[2] In one study of leks in European paper wasps, the aggregations of males lasted two to three months in the summer and recurred in the same locations year after year. Some resident males returned to the same landmark repeatedly in the course of the season; others appeared only transiently, visiting a number of leks. Resident males occupied their leks for an average of two weeks.[3] They were bigger than nonresident males, who used alternative mating tactics such "sneaking" into leks to secure mates.[4]

I suspect leks contribute to this wasp's urban success. Cities promote sexual isolation; by contrast, leks promote sexual mixing. Leks are explored further on page 70.

Chapter 6 // Bees, Wasps, and Ants

Sons from Unfertilized Eggs: cicada killer (*Sphecius speciosus*)

The female cicada killer produces sons by laying eggs that are unfertilized.

Cicada killers are solitary wasps, in contrast to yellowjackets, paper wasps, and hornets. Females are big, up to 40 millimeters (1.5 inches) in length, and 2.5 times heavier than males.[1] They are three to four times the size of yellowjackets.[2] A female cicada killer does all the work of digging and provisioning her nest. She stings and paralyzes a cicada and then carries it into a burrow she has dug in the ground. She places the cicada in a cell within the burrow and lays an egg on it. She provisions more cicadas in a cell destined to produce a daughter rather than a son. This sexual allocation requires that she know the sex of her eggs.[3]

Mated female cicada killers control the sex of an egg by controlling whether it is fertilized. To produce a son, a mother lays an egg that is unfertilized; it is haploid, having one set of chromosomes. To produce a daughter, she lays an egg that is fertilized; it is diploid, having two sets of chromosomes.[4]

Cicada killer broods contain three to four times more males than females. What explains this bias? Theory predicts that parental investment in sons and daughters will be equal.[5] Peter Grant, evolutionary biologist and ecologist at Princeton University, used cicada killer weight as a measure of parental investment. He concluded that higher weights of females compared to males supports this theory of equal parental investment.[6]

This cicada killer's haploid-diploid system of reproduction is typical of members of the order Hymenoptera, which includes bees, ants, and wasps.[8] Its adaptation of this system to cities is exquisite: a mother cicada killer methodically adjusts the size of her brood to availability of her prey.

Male and female cicada killers attracted to sap oozing from lilac in a city garden. The female is bigger.

Lopsided Sex Ratios: pavement ant (*Tetramorium immigrans*)

Reproductive members of some colonies of pavement ants are predominantly one sex.

Pavement ant.

Colony of pavement ants excavating sand between brick pavers.

Pavement ants specialize in populating cities, including buildings. They are the dominant species of ant in some urban locales.[1] Introduced from Europe, they have spread across North America.[2]

A pavement ant queen may mate with more than one male.[3] After mating, she establishes a nest and lays eggs. She initially produces workers—virgin females who forage and help raise her brood but do not lay eggs. Later she produces winged reproductives, both males and future queens. These reproductive ants leave the nest and mix and mate in swarms presumably with reproductive members of other colonies. After mating, males die; queens found new colonies and shed their wings. Each colony has only one queen.

Kenneth Bruder and Ayo Gupta, entomologists at Rutgers University, counted the number of reproductive females and males in eleven colonies of pavement ant in New Jersey. They found one colony produced seven times more males than reproductive females; another produced eleven times more reproductive females than males. Reproductive sex ratios in just under half the colonies were markedly skewed.[4]

Hypotheses that seek to explain skewed reproductive sex ratios in ant colonies typically focus on theoretical conflicts between queens and workers.[5] They fail to explain skewed reproductive sex ratios in ants generally. For example, studies that experimentally switched queens in fire ant colonies showed that the queen rather than her colony skewed the colony's reproductive sex ratio.[6]

Skewed reproductive sex ratios in colonies of the pavement ant are enigmatic in origin and function. The evolutionary pathways are likely complex.[7] As a practical matter, a markedly skewed reproductive sex ratio in an ant colony may discourage inbreeding. This ant's chemical recognition of nestmates might also reduce inbreeding.[8] Its polyandry likely supports genetic diversity. A recent genetic analysis found no evidence of inbreeding in North American pavement ants, in contrast to many other invasive ants.[9] The breeding system of the pavement ant appears well adapted to supporting genetic mixing in cities.

/ DRAGONFLIES AND PRAYING MANTISES /

European mantis (*Mantis religiosa*).

Philadelphia has four species of praying mantis, including three introduced from abroad. It has 46 species of dragonflies and damselflies (order Odonata).[1] Each species courts its own kind. Despite their success in recognizing mates, they exploit deception during sexual conflict. This chapter examines how female dragonflies and female mantises apply sexual mimicry both defensively and offensively. Sexual mimicry may benefit females, especially in cities.

Male Territoriality: blue dasher dragonfly (*Pachydiplax longipennis*)

Male blue dashers defend aquatic territory attractive to female blue dashers.

Wrap-around compound eyes provide vision in almost all directions.

The blue dasher is an aerial predator that uses its spiny legs to grab insects in midair.[1] It has a pair of compound eyes that wrap around both sides of its head. Each eye is composed of thousands of minute optical facets. The two eyes together provide a field of view that is almost a complete sphere.[2] They detect color across the visible spectrum plus ultraviolet.[3] They also detect polarization and motion.[4] In addition to the pair of compound eyes is a trio of small simple eyes (ocelli), which detect positional change and support aerial stabilization.[5]

Resting position.

The male blue dasher perches on a lookout overlooking quiet, shallow water with aquatic plants. He secures prospective mates by defending territory that attracts females looking for sites to lay their eggs. From his lookout he attacks prey, chases rivals, and pursues females. The area that he defends around his perch defines his territory. He attacks other species of dragonfly, even those larger than himself.[6] He signals intruders by pointing his abdomen up as a warning. An intruder avoids aerial combat by recognizing the warning and retreating.[7] Elevation of the abdomen may serve other functions as well: predator avoidance, thermoregulation, and stabilization.[8]

Warning position. This position also may have other functions.

One study discovered that about one in ten male blue dashers failed to secure a territory. Compared to a territorial male, a nonterritorial male mated as often or less than half as often, and his mates spent less than half as much time laying eggs.[9] Territoriality may reward male dragonflies, especially in cities where bodies of water suitable for laying eggs are scarce.

Chapter 7 // Dragonflies and Praying Mantises

Secondary Genitalia: blue dasher dragonfly (*Pachydiplax longipennis*)

A mating pair of blue dashers forms a characteristic heart-shaped wheel.

Photo by Nigel D. F. Grindley

Copulating pair of blue dashers. Male is on top.

The blue dasher mates in three steps, which begin before he meets his prospective mate.

First is translocation of semen. The male dragonfly excretes semen from his primary genitalia near the tip of his abdomen. He then bends his abdomen down and forward so that the semen on his abdominal tip touches accessory genitalia on the underside of his abdomen near his thorax. He charges his accessory genitalia with semen before he first mates, and he recharges after.

Second is clasping. The male flies over the female and hovers, aligning his abdominal tip over her head. Flapping her wings, she rises and presents her head. Using claspers at the tip of his abdomen, he holds onto the back of her head. The two then fly in tandem, with the male on top and in front, holding onto the female, who trails behind.

Third is copulation. While the two are connected in tandem in midair, or while the male is perched, the female pivots underneath the male, who is still clasping the back of her head with his abdominal pincers. She places her abdominal tip—her genitalia—against his accessory genitalia. Copulation lasts 10 to 40 seconds, but occasionally more than two minutes.[1]

After copulation, the male guards his mate. She proceeds to lay eggs on plants floating on the water or submerged just below the surface. He may hover over her and repel intruders, or he may keep watch from a nearby perch. Both sexes are likely to copulate again, either with each other or with new mates. One marked male was observed to copulate with 15 females.[2]

Female Mimicry of Males: blue dasher dragonfly (*Pachydiplax longipennis*)

Sexual mimicry may help female blue dasher dragonflies avoid harassment.

Mature female.

Mature male.

The male blue dasher may aggressively attempt to mate with a female blue dasher who is not his mate and who attempts to lay eggs in his territory. His sexual aggression against this female disturbs her egg laying. Her distinctive yellow markings broadcast her identity.

In 1962 Clifford Johnson, biologist at the New Mexico Institute of Mining and Technology, used painted models of perched dragonflies to test the accuracy of the male blue dasher's visual perception. He placed his models in territories of resident male blue dashers. Almost all male blue dashers displayed threats and flew directly at models painted to look like male blue dashers. No male blue dashers threatened models painted to look like females; on the contrary, almost all males attempted to mate with the female models. Johnson concluded that male blue dashers use blue coloration to recognize males, and yellow markings to recognize females. He noted that juvenile males have yellow markings like females, and that mature males treated juvenile males like females.[1]

Female mimicking a mature male.

Some female blue dashers mimic coloration of mature male blue dashers. Johnson evidently had not appreciated this. This mimicry (bottom right photo) presumably protects females from sexual harassment while they lay eggs. Such protection might serve them especially well in cities, where scarcity of ponds potentially concentrates males, increasing harassment. Female defenses against harassment are widespread in insects, and female mimicry of males is well documented.[2] An unanswered question in the case of the blue dasher is how female mimics manage to mate while masquerading as male. Conceivably, females turn blue only *after* they mate. Both sexes as juveniles have abdomens marked in yellow. Both sexes as they age turn blue. In females, the transformation to blue is slower.[3] Perhaps mimicry of males is a late phase in the life of females.[4]

　　Chapter 7 // Dragonflies and Praying Mantises

Polarized Light Pollution: dragonflies (Odonata)

Polarization of light reflected off buildings and manufactured objects disturbs sexual behavior of aquatic insects.

Flying aquatic insects orient in part by detection of polarization of light reflected off water. They confuse this polarization with polarization of light reflected off glass and metal.[1]

Hansruedi Wildermuth and Gâbor Horváth at the University of Zurich described a male dragonfly that guarded a stationary car like a pond. He perched on the car's radio antenna and faced his "territory." He defended this territory from rivals. He periodically took off, flew in loops and circles, and returned to his perch. He darted after a second male that approached the vehicle. The two contestants spiraled up and away from the vehicle and out of sight, but one of the two soon returned to the antenna. When the investigators pushed in the antenna and replaced it with a stick, he adopted the stick as his perch. When they slowly moved the car, the dragonfly hovered over it and retained his lookout on the stick.

Investigation of polarization determined that light reflected off the car was horizontally polarized, and that this polarization continued across the visible spectrum, from blue to red.[2] Previous research had shown that horizontal polarization of light reflected off water is a cue that female dragonflies use for selecting aquatic sites for laying eggs.[3] Dragonflies have been observed laying eggs on the windshields of cars and on slowly moving shiny wheels of a truck.[4] Black tombstones reflecting horizontally polarized light in cemeteries have trapped dragonflies attracted to them.[5]

Reflective glass façade of Cira Center, along Schuylkill River, Philadelphia.

Diverse kinds of polarizing reflective surfaces attract insects. They include plastic sheeting, windowpanes, solar panels, asphalt, and crude oil. In addition to dragonflies (Odonata), aquatic insects attracted to polarizing reflective surfaces include caddisflies (Trichoptera), stoneflies (Plecoptera), mayflies (Ephemeroptera), true bugs (Hemiptera), flies (Diptera), and beetles (Coleoptera).[6]

In spring in downtown Philadelphia, dead dragonflies have littered sidewalks next to skyscrapers clad in glass. Perhaps the insects mistook the glass for places to mate or lay eggs.

Female swamp darner (*Epiaeschna heros*) found dead on sidewalk beside skyscraper clad in glass.

Monandry: fragile forktail damselfly (*Ischnura posita*)

The female fragile forktail mates once.

Mature female.

Mature male

Damselflies are aerial predators in the same insect order as dragonflies. Larvae of both are aquatic predators. While the female blue dasher dragonfly mates multiple times, the female fragile forktail damselfly apparently mates only once.

Mating in this damselfly is so rare that it is easy to overlook. In studying this damselfly, James Robinson, an evolutionary ecologist and an aquatic insect specialist at the University of Texas, marked 2,215 individuals and witnessed none mating. He observed females rejecting males attempting to mate with them. When a suitor approached an unreceptive female, she spread her wings and curled the tip of her abdomen downward.[1]

Young female fragile forktail damselflies have bright metallic blue markings that distinguish them from older females and from males. Older female fragile forktails are dull grayish blue or brown, but unlike older female blue dashers, they do not mimic males; males retain their brilliant stripes of metallic green. Two rare photos documenting mating in fragile forktail damselflies show that the coloring of copulating females is dull.[2] Perhaps dull coloration protects female fragile forktail damselflies from predators. This defense may outweigh advantages of bright, attractive sexual advertisement.

Predators that attack young, brightly colored females include older female fragile forktail damselflies. Robinson observed three cases of these older females cannibalizing young vulnerable females.[3] Cannibalism nutritionally benefits the cannibal, and it may relieve competitive pressure on her offspring for food and space. This benefit theoretically might apply especially in cities, where ponds suitable for breeding are scarce.

Chapter 7 // Dragonflies and Praying Mantises

Sexual Cannibalism: praying mantises (Family Mantidae)

Female praying mantises cannibalize male praying mantises.

Female praying mantises in the wild cannibalize males. Lawrence Hurd, biologist specializing in predatory arthropods at the University of Delaware, and colleagues found that sex ratios of praying mantises shifted to predominantly female toward the end of the season. Most of the time when females ate males, they ate males as prey, without copulation. And most of the prey that the researchers observed female praying mantises eat were male praying mantises.[1]

Before this discovery of predation, researchers thought that praying mantis females ate males primarily after mating with them, or they viewed cannibalism as a mere artifact of captivity with males and females crowded together. Cannibalism during mating theoretically serves the male's immediate reproductive interests: It nutritionally supports the mother bearing his offspring—unless his mate is polyandrous.[2] By contrast, cannibalism simply as predation denies the male victim all future opportunity to pass on his genes.

Gravid female Carolina mantis (*Stagmomantis carolina*) eating a painted lady butterfly (*Vanessa cardui*) on butterfly bush (*Buddleia*) in a city garden. The mantis's wings are too small for flight.

Female on left, male on right. The male's eyes are bigger than the female's eyes. In addition to the pair of big compound eyes is a trio of tiny amber-colored simple eyes, located in the middle of the head. Males visually monitor females during courtship.[3] The species is the Chinese mantis (*Tenodera sinensis*).

Femme Fatale: praying mantises (Family Mantidae)

Some female praying mantises lure males for food rather than for sex.

Male Carolina mantis in the same rowhouse garden as the female Carolina mantis shown on the preceding page. His abdomen is slender compared to hers. His wings are longer and can power flight over short distances. He disappeared from this garden when the female appeared, but it is unknown whether he emigrated, was cannibalized, or vanished for some other reason.

Sex pheromone of female praying mantises attracts males over long distances, reportedly up to 100 meters.[1] Katherine Barry of Macquarie University in Sydney investigated how female praying mantises use sex pheromone to transform into "femmes fatales," cannibalizing their suitors. She found that better-fed and more fecund female mantises attracted more males. This suggested that higher production of female sex pheromone advertised superior fitness and fecundity.

Barry went on to investigate whether female mantises exploit sex pheromone deceptively. She methodically varied food supplied to virgin females (of the false garden mantis, *Pseudomantis albofimbriata*). As expected, better-fed females were more fecund and attracted more males, but with one notable exception: females that Barry fed the least. These malnourished females had the lowest fecundity but paradoxically attracted the most males. Barry concluded that for these hungry females, sex pheromone functioned as a deceptive lure, falsely advertising high fecundity. In controlled trials, starved females ate their suitors 90 percent of the time, compared to zero percent for well-fed females; the same starved females copulated with their suitors only 50 percent of the time compared to 100 percent for well-fed females. Starved females transformed into seductive cannibals, or femmes fatales.[2]

When abundance of prey wanes in late summer in city gardens in downtown Philadelphia, male Carolina mantises (*Stagmomantis carolina*) can fly to more rewarding hunting grounds. By contrast, female Carolina mantises cannot fly. Although they can crawl, they are largely stuck where they happen to be at the end of the season. For these starving females, the only means for securing food may be to make food come to them—that is, to transform into femmes fatales.[3] In studies of mature Carolina mantises in captivity, females cannibalized males, but males never cannibalized females.[4]

8

/ OTHER INSECTS /

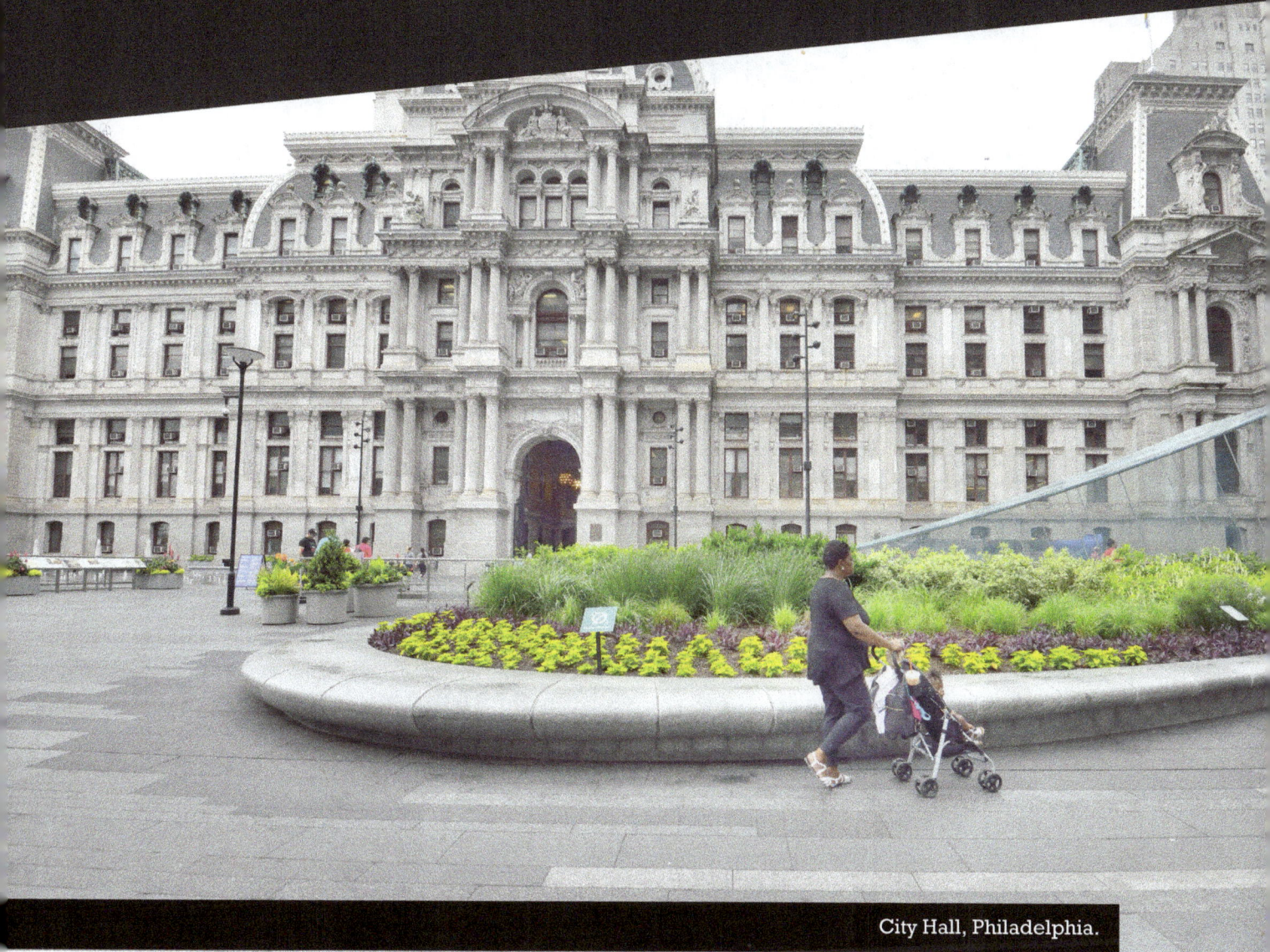

City Hall, Philadelphia.

In 2019, Philadelphia's City Council passed a bedbug ordinance. It codified when landlords are responsible for costs of interventions for control of bedbugs.[1] The ordinance is testimony to the reproductive power of this pest in Philadelphia. Other introduced insect pests that proliferate in the city include the northern house mosquito and the oriental cockroach. Some native insects such as the common eastern firefly also find the city hospitable. By contrast, the locust borer, another native, has lost crucial places to meet and mate downtown. This chapter considers how mating systems of exotic and native insects downtown have endured.

Femme Fatale: fireflies (Family Lampyridae)

Male fireflies emit coded flashes exploited by predatory females of other species of firefly.

Copulating pair of common eastern fireflies.

Mating pair after dismounting.

The common eastern firefly (*Photinus pyralis*) is native to eastern North America. The male flashes to signal females. He begins flashing about a meter off the ground, in a pattern specific to his sex and species. Resting on a blade of grass below him, a female common eastern firefly recognizes his flash code. After a delay of several seconds, she answers with her species-specific flash code. He recognizes her code and drops down. As they engage in a flash dialogue, he approaches her, and they mate.[1]

Photuris fireflies are also native to eastern North America. They too flash to signal mates, but after they mate, females of some *Photuris* species transform into femmes fatales.[2] These femmes fatales prey on males of other species of firefly, like the common eastern firefly. They recognize their prey's flash codes and answer with flash codes that mimic those of their prey's prospective mates. As a conned male approaches his anticipated "mate," his femme fatale seizes him and eats him alive.[3]

Femme fatale fireflies have repertoires of fake codes matching those of females of different species of prey. They recognize the male codes of their prey and customize their deceptive answering flashes accordingly.[4] *Photuris* femmes fatales sometimes cannibalize males of their own species.[5]

In studies investigating induction of flashing at night, *Photuris* fireflies required background darkness two orders of magnitude darker than that required by the common eastern firefly.[6] In downtown Philadelphia, I have yet to find *Photuris* fireflies, but the common eastern firefly is abundant. The common eastern firefly begins flashing at dusk. Conceivably, light pollution downtown repels *Photuris* fireflies and affords the common eastern firefly a safe haven.

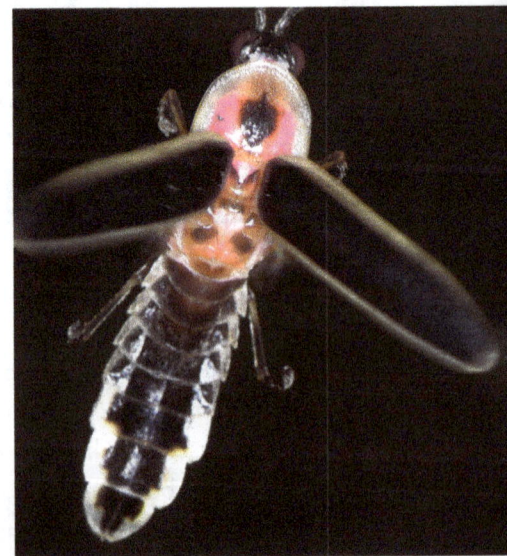
Airborne male projecting his light down. The camera is above him.

Contest Competition in Males: reddish-brown stag beetle (*Lucanus capreolus*)

Male stag beetles use their giant mandibles in contests with other male stag beetles.

The reddish-brown stag beetle is native to eastern North America. Stag beetles get their name from their big mandibles, which in some species are ornate like antlers. The mandibles of the reddish-brown stag beetle are more modest, although still formidable looking. Despite their appearance, the male's bite is not painful to people.

Two theories have been advanced for the function of these mandibles. The first attributes their use primarily to fights with other males over females. Males have been reported to joust with each other for dominance as females wait nearby. Females

Male reddish-brown stag beetle attracted to the illuminated sheet shown on page xxvii. On the right, an ant provides a scale for size.

choose the victor as mate. In captivity, males placed near each other fought, but the rate of attacks immediately increased if a female was placed nearby. The sight of another male appeared to trigger attacks.[1]

A second theory focuses on competition for resources, such as tree sap, which attracts females. Males mate with females that come to feed on sap. According to this theory, the proximate cause for fighting between males is competition for sap and not females.[2] Documentation of male mating contests outside of captivity is sparse in the reddish-brown stag beetle. In the best-documented case, male fighting was attributed to competition over females, but alternative hypotheses were not considered.[3]

Male mandibles also function in courtship. The male presses his mandibles against the female after he mounts her. During copulation, he thrusts his mandibles forward. He overpowers her when she attempts to flee.[4]

Male-male conflict in this stag beetle favors size and strength, in contrast to other forms of male-male sexual competition mentioned in this book. For example, size and strength are not advantageous in scramble competition in the eastern gray squirrel, in competition among male orange sulfurs, and in cuckoldry in bluegills.

Vulnerable Mating Place: locust borer (*Megacyllene robiniae*)

The locust borer beetle meets and mates on flowers of goldenrod (Solidago spp.).

Locust borers mating on goldenrod.

Female locust borer feeding on tall boneset.

The locust borer beetle is native to eastern North America. Its range has expanded with that of its larval host plant, black locust (*Robinia pseudoacacia*), a tree native to the same region but introduced widely in the United States.[1]

Jimmy Galford of the U.S. Forest Service used caged virgin female locust borers as bait to attract males. This attempt failed, and he concluded that females do not attract males from a distance. He then placed males in an aquarium with twigs, some previously exposed to virgin females. The males fought with each other for positions on the female-exposed twigs. Galford concluded that females deposit on their substrate a sex pheromone that attracts males.[2] Later, a contact sex pheromone from the cuticle of females was chemically identified.[3]

For the locust borer, goldenrod functions not only as a source of food, but also as a place to meet and mate.[4] Goldenrod and black locust are both pioneer species, colonizing areas recently cleared of vegetation. The two plants occur together along roadsides and railroads. The locust borer beetle feeds on flowers of other species such as tall boneset (*Eupatorium altissimum*), but I have seen it mating exclusively on goldenrod. In downtown Philadelphia, mugwort has largely replaced goldenrod. Downtown, this big, colorful beetle has disappeared from its old mating areas even though its larval host plant, black locust, endures.

Goldenrod (*Solidago* sp.) growing with black locust (*Robinia pseudoacacia*).

Sexual Emigration: red milkweed beetle (*Tetraopes tetrophthalmus*)

Male red milkweed beetles emigrate when they outnumber females.

Small male on top of large female red milkweed beetle. They are on common milkweed (*Asclepias syriaca*).

The red milkweed beetle is native to eastern North America. The beetle feeds on milkweed foliage while its larva feeds on milkweed roots. Female milkweed beetles puncture midribs of milkweed leaves and drain their toxic latex. Male milkweed beetles feed preferentially on leaves that female milkweed beetles have processed.[1]

In studies, males flew preferentially to milkweed patches that had many big flower heads (as shown on page 19). They landed on plants independent of the presence of females.[2] Both males and females mated many times a day. Males often forced mating pairs to disengage.[3] Larger males won contests for mates, and males preferentially mated with large females.[4] Male milkweed beetles moved twice as far as females within milkweed patches and emigrated more often to other patches.[5] Males tended to fly and emigrate when they outnumbered females.[6]

Theoretically, emigration of males to alternate milkweed patches promotes genetic mixing. It may also promote mating, depending on the availability of females in alternate patches. Dispersal of the red milkweed beetle across milkweed patches in cities may help the species overcome genetic isolation due to fragmentation of habitat.

Leks: northern house mosquito (*Culex pipiens*)

The northern house mosquito mates above ground and below ground, but above ground it mates in leks. These leks function like those of the European paper wasp.

Larva of a *Culex* mosquito, probably the northern house mosquito. I collected it from our birdbath. An air tube extends up to the surface.

The northern house mosquito is native to Africa but is common across temperate North America, Europe, and Asia.[1] It mates in swarms, or "leks." One report from Urbana, Illinois, over a century ago described each swarm as aligned above a tall object, such as a telephone pole, tree, or shrub. A typical swarm measured about half a meter in diameter and made a characteristic, high-pitched sound. A female would fly in from nearby foliage, dart into a swarm, and emerge copulating, dragging her mate end-to-end in midair. As darkness came, the swarms dispersed. One swarm included 897 males and 4 females.[2]

Leks in the northern house mosquito engage members of populations that mate above ground. But some populations of this species complete their entire life cycle below ground. Males below ground do not mate in leks but rather settle on females at rest.[3] Subterranean populations have been reported in London, New York, Philadelphia, Boston, Chicago, and Baltimore.[4] They occur in sewers and subway tunnels. In New York City and London, gene flow between populations below ground and above ground was found to be minimal or absent.[5]

Above-ground populations express three additional traits that distinguish them from below-ground populations. Above-ground adult females (1) overwinter in an inactive state, (2) feed predominantly on birds, and (3) require a blood meal to lay eggs. By contrast, below-ground adult females (1) remain active throughout winter, (2) feed predominantly on mammals, and (3) do not require a blood meal to lay eggs.[6]

The northern house mosquito illustrates how cities can support extreme sexual isolation even in a species that is widespread and abundant. The above-ground and below-ground populations of the northern house mosquito were once regarded as separate species. Currently they are viewed as two forms (or biotypes) of the same species.[7] In southern Europe members of the two populations occasionally occur together and hybridize.[8]

Facultative Parthenogenesis: oriental cockroach (*Blatta orientalis*)

A female oriental cockroach can reproduce despite failure to find a mate.

Female oriental cockroach.

Male oriental cockroach.

Despite its name, the oriental cockroach is native to North Africa. It was introduced into North America from Europe.[1] Archeologists in England discovered remains of this species in urban deposits dating back to the fourth century.[2]

In studies of the German cockroach (*Blattella germanica*), siblings preferred each other as social partners rather than sexual partners. They also preferred the company of full siblings over half siblings or cousins. Hydrocarbons in cuticles provided genetic cues that helped siblings

Juvenile oriental cockroaches, socializing.

avoid mating with one another. Chemical differences correlated inversely with relatedness. Cockroaches benefited from avoiding mating with siblings: mated siblings produced embryos with increased rates of abortion.[3]

To avoid inbreeding, oriental cockroaches disperse; but dispersal may deprive them of access to mates. Female oriental cockroaches emit volatile sex pheromones, and some evidence suggests that males emit sex pheromones as well.[4] But neither sex can fly; wings of female oriental cockroaches are reduced to mere stubs, and wings of males, although bigger, are too small for flight. After they disperse, females may fail to attract prospective mates.

A virgin female oriental cockroach that fails to find a mate can lay viable eggs parthenogenetically (without fertilization); however, her eggs produce only daughters, and these daughters must find mates in order to reproduce.[5] By contrast, American cockroaches escape this limitation. A unisexual colony of virgin female American cockroaches (*Periplaneta americana*) propagated parthenogenetically over many generations for three years.[6] Virgin female German cockroaches are incapable of parthenogenesis; they can produce eggs, but they fail to hatch.[7]

Parthenogenesis: milkweed aphid (*Aphis nerii*)

Milkweed aphids reproduce by parthenogenesis; they make offspring from unfertilized eggs.

Colony of milkweed aphids: all female.

A milkweed leaf (*Asclepias syriaca*) with aphids.

Milkweed aphids (also known as oleander aphids) are native to the Mediterranean region. They bear their young live. They produce no males and never mate. Their eggs develop into embryos without fertilization. They eliminate courtship, sexual harassment, mate competition, sexually transmitted disease, and paternal genetic disorders. They eliminate risks of reproductive failure due to lack of availability of mates. Perhaps most importantly, they eliminate the cost of producing sons—who bear no young. They clone (preserve) their genotype.[1] They also telescope generations: a milkweed aphid bearing young carries her daughters and embryonic granddaughters. At birth, her daughter is already pregnant. Parthenogenesis makes telescoping of generations possible. It enables eggs to begin embryogenesis before their mother is born. By contrast, in sexual reproduction, embryo development occurs only after the mother has mated, which can happen only after she is born.[2]

Winged milkweed aphid near immatures.

A lone virgin female milkweed aphid can quickly establish a colony. Her mode of reproduction is well adapted to populating isolated, ephemeral urban habitats.

Chapter 8 // Other Insects

Cyclic Parthenogenesis: poplar leaf aphid (*Chaitophorus populicola*)

The poplar leaf aphid alternates between reproduction with and without fertilization.

Young eastern cottonwood tree (*Populus deltoides*).

Ants tending aphids on eastern cottonwood.

The poplar leaf aphid is native to North America and distributed widely across the continent.[1] In spring, female poplar leaf aphids hatch from eggs. These virgin females bear their daughters live. Generations of virgin daughters continue giving live birth parthenogenetically until fall, when they produce sons and daughters. Aphids from this last generation of the season now mate and lay fertilized eggs that overwinter. In the spring, the eggs hatch into females, continuing the cycle.[2]

In response to shortening day length in the fall, some ova switch to having only one, rather than two, X chromosomes. Ova with two X chromosomes develop into daughters—maternal clones. Ova with one X chromosome develop into sons.[3] Cyclic parthenogenesis enables the poplar leaf aphid to reap benefits of both sexual reproduction and cloning. Jean-Christophe Simon at the French National Institute of Agricultural Research, and colleagues, concluded that a major benefit of sexual reproduction in aphids is production of cold-tolerant eggs.[4] In Philadelphia, sexually produced eggs enable the poplar leaf aphid to overwinter. By contrast, the strictly parthenogenetic milkweed aphid lays no eggs. It bears its young alive. It cannot overwinter this far north. Every spring it has to migrate into this city from the south and establish new populations.

Live birth of poplar leaf aphid on eastern cottonwood.

Traumatic Insemination: bedbug (*Cimex lectularius*)

Traumatic insemination of female bedbugs injures females but may benefit the species.

Bedbug crawling off a penny.

The male bedbug jabs his intromittent organ through his mate's abdominal wall rather than inserting it through her vaginal opening. The abdominal perforation creates an open wound.[1] He times his attack to occur after she has had a blood meal, which slows her down.[2] He engages in no courtship. He will mount any suitably sized object, even a piece of cork.[3] To protect themselves from attack, male and immature bedbugs produce anti-aphrodisiac and other defensive pheromones.[4]

The last male to mate with a female bedbug before she lays eggs is the male most likely to sire her offspring. This encourages males to mate often. In studies, females experienced traumatic insemination on average five times within a period of less than two hours after a blood meal. The number of eggs a female laid after a blood meal was the same regardless of whether she mated once or more than once. By contrast, females that mated more than once incurred increased mortality, and they produced fewer eggs over the course of their lives.[5]

Males inject their semen into the female through a specialized site on the abdominal wall into an organ dedicated to receiving sperm. This organ protects females against wound complications, such as infection and dehydration.[6] Females may extract nutritional value from ejaculate, which experimentally increased rate and duration of reproduction.[7]

Margie Pfiester and colleagues at the University of Florida noted that traumatic insemination induces mated female bedbugs to avoid males. They hypothesized that it encourages dispersal, serving the interests of the species.[8] The bedbug was introduced to North America from the Old World.[9] Local dispersal of bedbugs to people and their belongings has helped to establish the bedbug as a pest worldwide.[10]

Traumatic insemination in animals has evolved independently at least 36 times.[11] It includes members of six orders of insects and three phyla of worms, and the giant squid. Species of affected insects include certain bedbugs, plant bugs, ants, moths, flies, beetles, and crickets.[12]

9

/ INVERTEBRATES EXCLUSIVE OF INSECTS /

Bridge spider (*Larinioides sclopetarius*) illuminated in her web at a municipal lamp at night.

Bright yellow markings of the female bridge spider practically frame her genitalia. In related nocturnal species, these markings have been shown to attract prey. Whether they function also to attract mates, or to assist in copulation or sexual cannibalism, is unknown.[1] In spiders as in praying mantises, sexual conflict has encouraged females to cannibalize males.[2] Sexual conflict in spiders is expressed also in female genital mutilation—a means by which males prevent their mates from copulating with other males.[3] This chapter considers how vulnerabilities tied to mating systems in cities may shape invertebrate populations.

Sexual Mixing Around Outdoor Lights: bridge spider (*Larinioides sclopetarius*)

Outdoor lighting interferes with nocturnal mating and pollination by moths, but it facilitates mating by bridge spiders.

Egg sacs of bridge spiders. The egg sacs are gray or white spheres covered with silk, superficially resembling cotton balls. They are clustered around a flood light and under an eave, which shields them from rain.

Native to Europe and Asia, bridge spiders in North America range from Alaska and Newfoundland south to North Carolina.[1] They spin vertical orb webs near electric lights close to bodies of water. They feed on flying aquatic insects, especially midges, attracted at night to the lights. Sexually mature males do not construct webs; they capture prey in webs belonging to females.[2]

Under experimental conditions, sexually mature female bridge spiders constructed their webs preferentially at sites that were artificially illuminated. Artificial light promotes establishment of populations of bridge spiders in high density.[3] At Cincinnati's illuminated Riverfront Coliseum Sports Arena, bridge spiders constructed communal webs with as many as 100 spiders per square meter.[4]

Artificial light functions as a place where bridge spiders eat, meet, and mate. Such a place is shown in the photo. Here, bridge spiders have propagated downtown around a flood lamp under an eave of a building beside a stagnant inlet of the Schuylkill River. The spiders have concentrated their egg sacs close to the light. The light will supply their offspring with nocturnal flying insects emerging from the river.

Chapter 9 // Invertebrates Exclusive of Insects

Two Sets of Male Genitalia: bridge spider (*Larinioides sclopetarius*)

Male spiders transfer sperm from primary to secondary genitalia before mating.

Male, with spherical enlargement of pedipalps in front of head.

Female, without spherical enlargement of pedipalps.

The process by which male spiders transfer sperm to their mates is unique to spiders. The first time a male spider discharges semen, the recipient is not the female. He emits a drop of semen initially onto a strand or web of silk. He then applies to this semen his pedipalps—a pair of appendages extending in front of his head. The end of each pedipalp is spherical and functions as a storage vesicle for sperm (left photo). He draws up semen into these seminal vesicles, which hold it until he mates.[1]

During mating, pedipalps both store and deliver sperm. The bridge spider places one of his pedipalps onto female structures overlying paired female genital openings on the underside of her abdomen. Pincers on his pedipalps grip a projection from her genitalia.[2] Bridge spiders may mate three or four times with each pedipalp.[3] They recharge their pedipalps with sperm by repeating the procedure they used to fill them initially.[4] After mating, female orb weavers store semen until they are ready to lay their eggs. Sperm fertilizes their eggs as they lay them.[5]

All spiders have the same basic system of transfer of semen to pedipalps and to female genitalia. Differences in morphology of genitalia distinguish species and presumably help to keep different species reproductively isolated from each other. The genital bulb that stores semen in male pedipalps lacks neurons and sensory structures.[6] Theoretically, when this bulb is in the field of view of male eyes, male vision could assist in guiding copulation. Perhaps artificial lighting at night helps bridge spiders copulate. This hypothesis has yet to be explored.

Although only spiders transfer semen to pedipalps, other arachnids and some insects have secondary genitalia. Function of secondary genitalia in dragonflies is described on page 59.

Sexually Compatible Personalities: bridge spider (*Larinioides sclopetarius*)

At lamps, potentially cannibalistic bridge spiders live and mate close together.

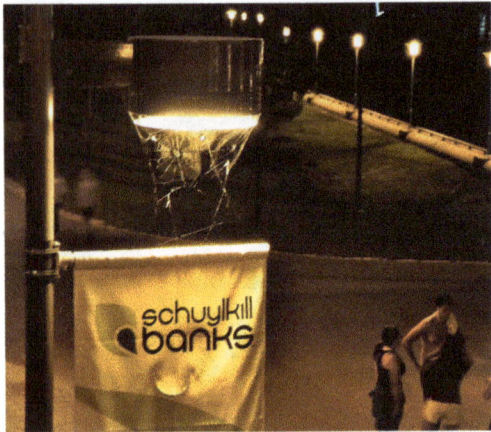

Webs clustered together under a lamp on a riverbank.

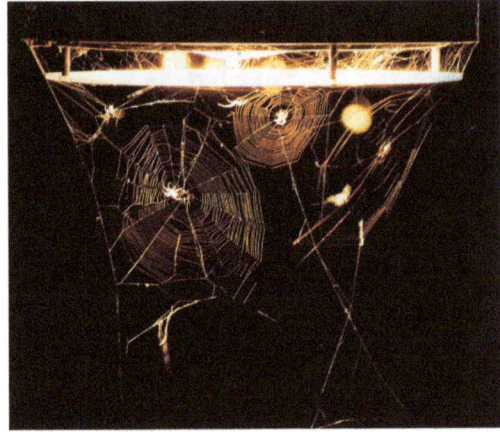

Close-up of webs shown in photo on left.

Bridge spider courtship is potentially dangerous. Both sexes are predators, and the goal of each is to mate without being eaten. Sexually mature male bridge spiders do not make webs, but rather wander about searching for females. They search at night when darkness impairs vision. Their prospective mates are on average bigger, and at night these females are waiting to seize prey in the same webs that males enter to mate. A bridge spider waiting for prey in a web responds aggressively in 0.1 second to a vibration with an amplitude of only 10 microns (0.0004 inch).[1]

In studies, a male suitor in the web of a western black widow (*Latrodectus hesperus*) avoided provoking an attack. His vibratory signals were low in amplitude compared to high-amplitude vibrations of potential prey. They were also continuous and of long duration compared to the short, sporadic, percussive vibration of snared insects.[2] A male suitor in the web of a labyrinth orb weaver (*Metepeira labyrinthea*) initiated a vibratory dialogue with his prospective mate. He alternately advanced and retreated. She kept her position in her web, occasionally plucking the web lines with her front legs.[3] Sex pheromones may reduce dangers of courtship.[4] Male bridge spiders have courted females outside orb webs.[5]

Simona Kralj-Fišer and colleagues at the Slovenian Academy of Sciences and Arts found that docile male bridge spiders were more likely to mate with docile females, and that aggressive males were more likely to mate with aggressive females.[6] Aggressiveness or docility in any individual spider over time stayed constant, suggesting that personality traits were genetic. Populations of urban spiders contained a mixture of both personality traits. The investigators theorized that natural selection in a colony regulates the proportion of spiders with each type of personality.[7] One might hypothesize that mortality of spiders with aggressive personalities goes up when these spiders concentrate around a lamp. Such mortality could explain the peaceful coexistence of bridge spiders close together at a lamp, as shown in the photos at the top of the page.

Genetic Mixing in the City: house spiders (Araneae)

Cities genetically isolate some populations but genetically mix others, like those of house spiders.

Yellow sac spider (*Cheiracanthium* sp.) photographed on a wall inside a house in downtown Philadelphia.

The western black widow spider (*Latrodectus hesperus*) is a poisonous urban house spider native to the western United States. Its young disperse by ballooning on strands of silk.[1] Western black widow spiderlings readily pass through window screens.[2] Individuals of this species have hitchhiked to South Korea and Maine.[3]

Gene flow between two populations of spiders occurs when spiders from one population mate with spiders of the other. Lindsay Miles at Virginia Commonwealth University and colleagues applied molecular genetics to track gene flow of the western black widow. They found that gene flow within cities was higher than gene flow outside cities, and that gene flow was higher between cities than between areas outside cities. Urban populations acted as hubs, connecting isolated populations outside cities.[4]

The yellow sac spider shown in the photo is probably *Cheiracanthium mildei*, a house spider whose distribution is even wider than that of the western black widow. Introduced from Europe, it is established in eastern and western North America and also in Africa, Asia, and South America.[5] It bites human beings.[6] House spiders illustrate the importance of cities as genetic mixers of small urban organisms, including many species presented in this book.

Mating Without Copulation: house centipede (*Scutigera coleoptrata*)

The male house centipede deposits on the ground a container filled with his sperm.

Male house centipede on top of a small moth that he is eating. The head of the centipede is on the right.

Most terrestrial animals transfer sperm to their mates by direct contact. House centipedes are an exception. These fanged predators eat common household insects, including flies, silverfish, moths, and cockroaches.[1] Prospective mates court by touching each other with vibrating antennae. The male eases himself under his mate and begins to rock. He then deposits on the ground a container filled with his sperm. It is almost half a centimeter in length. The outer wall of the container is incomplete, affording the female access to its contents. He leads her to the container, and she uses the tip of her abdomen to take in his sperm.[2]

Red-back salamander in a park in downtown Philadelphia.

Other terrestrial animals transfer sperm without copulation. They include certain species of scorpion, mite, springtail, and millipede, plus other centipedes.[3] Some terrestrial vertebrates have independently evolved indirect transfer of sperm. The male red-back salamander (*Plethodon cinereus*) deposits his sperm-filled container typically on a fallen leaf. The female puts the container into her cloaca.[4] Other amphibians, such as the American toad (page 89), fertilize eggs externally in water as the female spawns. Evolution of fertilization on land helped to liberate ancestors of terrestrial salamanders from dependence on water. It also preadapted them to cities.

Chapter 9 // Invertebrates Exclusive of Insects

Androgenesis: Asian clam (*Corbicula fluminea*)

Inheritance in Asian clams is almost exclusively paternal.

Site of riverbed covered with Asian clam shells.

Asian clams threatened to clog cooling water intake pipes to this power plant.[6]

Asian clams are hermaphrodites. Male and female sexes reach maturity at the same time, enabling "self-fertilization."[1] After sperm from an Asian clam penetrates an Asian clam egg cell, the egg cell ejects material containing its own nuclear genes. The embryo inherits all its nuclear genes exclusively from sperm. Only a tiny fraction of the embryo's total genetic material is maternal. This material lies in mitochondria, small bodies inside cells but outside the cell nucleus where the bulk of DNA is stored. This rare quasi-sexual form of reproduction is termed androgenesis.[2]

Asian clams are practically paternal clones. Lise-Marie Pigneur at the University of Namur, Belgium, and colleagues called them "selfish shellfish" because the male gamete acts like a selfish gene. The male gamete passes all his genes on to offspring, while the female gamete passes on almost none of hers.[3] The Asian clam has invaded rivers and streams across the United States. A lone clam can establish a colony. But the species fails to generate genetic diversity. It is vulnerable to environmental stress, such as temperature extremes and low oxygenation.[4] It is prone to mass mortality. Putrid die-offs have occurred widely, including in Philadelphia.[5]

Asian clam shells on a riverbank. (Penny added for scale.)

Facultative Self-Fertilization: gray garden slug (*Deroceras reticulatum*)

Like the oriental cockroach, the gray garden slug can reproduce despite failure to find a mate.

Gray garden slug with mucous trail.

Mating pair with a third slug in attendance.

Habitat of the gray garden slug.

Gray garden slugs track mucous trails of prospective mates. In one study, gray garden slugs crossing mucous trails turned almost immediately and faced in the direction of the slugs that had laid the trails. They then proceeded to follow the trails in the correct direction; they tracked trails as old as 6.5 hours.[1]

After a gray garden slug tracks down its mate, the tracker's head meets the mate's tail. If the two are sexually attracted, they perform a "nuptial dance." They form a circle, with the head of each facing the tail of the other. The circle rotates clockwise on a bed of mucus. As the circular pair rotate, the two progressively come closer together until their genital openings align, producing a yin-yang configuration. Each caresses the other with a conical erectile organ (sarcobelum) that protrudes from a location near its front end. The nuptial dance lasts 30 minutes to 2.5 hours and may attract other gray garden slugs. Transmission of sperm in the gray garden slug (a hermaphrodite) is reciprocal, meaning each member of a mating pair passes on sperm to the other.[2]

If a gray garden slug fails to mate, it can nevertheless lay eggs that hatch. It probably self-fertilizes. In studies, self-fertilized eggs developed more slowly and hatched at a lower rate than cross-fertilized eggs. Self-fertilization produced offspring for only one generation. Slugs that were the product of self-fertilization could not reproduce by self-fertilization; they laid eggs that failed to hatch.[3] Whether gray garden slugs produced by self-fertilization can propagate by cross-fertilization is unknown. If they can, self-fertilization for one generation followed by cross-fertilization the next could prevent reproductive failure.

Sexual Conflict of Interest: nightcrawler earthworm (*Lumbricus terrestris*)

Each nightcrawler in a mating pair has a reciprocal interest in injuring the other.

This earthworm (*Octolasion tyrtaeum*) avoids injury that mating nightcrawlers inflict on each other. Unlike nightcrawlers, it reproduces parthenogenetically (without mating).

Nightcrawlers are hermaphrodites. Each reciprocally inseminates the other. Mating earthworms face each other in opposite directions on the soil surface. Sexual structures on their forward sections are aligned in a way that allows sperm released by one worm to pass (through a circuitous route) into sperm receptacles (spermathecae) of its mate.[1]

The nightcrawler uses 40 to 44 needle-like bristles to perforate the skin of its mate during copulation. These bristles do more than hold the pair together. In a study of copulation, they were shown to cause tissue damage and to inject glandular secretions.[2] The tissue damage to skin of a mate during copulation may compel the mate to delay remating with another earthworm.[3] This delay potentially benefits the inseminator: a nightcrawler can fertilize its eggs with sperm stored for up to six months after it has been inseminated.[4] A nightcrawler that forces its mate to delay remating reduces the likelihood that sperm from a future rival will outcompete its sperm for fertilization of its mate's eggs.[5]

Each nightcrawler in a mating pair injures the other to serve its own reproductive interests, but the sexual conflict goes only so far. Injury to a mate would be counterproductive if it caused death or prevented egg laying. Other animals that injure their mates to discourage them from remating include insects, spiders, mammals, and birds.[6]

Lethal Mating Trails: terrestrial flatworms (*Bipalium* spp.)

Predatory flatworms track down nightcrawlers (L. terrestris) by following their trails.

Plant nurseries have incidentally introduced Australian and Asian flatworms (planarians) into Europe and North America. Several species of these introduced flatworms are specialized predators of earthworms.[1] They locate their prey by following their trails. Their prey includes nightcrawlers, which leave mating trails that are exceptionally long—up to 70 centimeters (28 inches). Earthworms leave trails also when they forage and disperse.[2]

Terrestrial flatworm (*Bipalium pennsylvanicum*).

Cara Fiore and colleagues at the State University of New York College at Cortland observed flatworms (*Bipalium adventitium*) hunting nightcrawlers and other earthworms in the laboratory. After a flatworm tracked down its prey, it mounted it, opened its mouth, everted its pharynx, attached its pharynx to the worm, secreted digestive enzymes, and liquified and ingested it. Most species of earthworm in North America are native or European in origin and have had no evolutionary opportunity to evolve defenses against flatworms from Asia and Australia.[3]

After introduction of an exotic predatory flatworm into Belfast, Northern Ireland, a parthenogenetic species of earthworm increased in abundance while six other species of earthworm either declined in abundance or disappeared. Nightcrawlers were among those that disappeared. Parthenogenesis may have protected the species of earthworm whose abundance increased. Parthenogenetic earthworms do not leave mating trails (although they do leave other trails that flatworms may follow). In addition, a lone parthenogenetic earthworm can restore an earthworm population that flatworms have driven nearly to extinction.[4]

An open question is whether proliferation of predatory flatworms in North America will shift communities of earthworms toward species that are parthenogenetic. Dissemination of predatory flatworms may shift these communities also toward earthworms introduced from Asia. One Asian earthworm (*Amynthas agrestis*) widely established in North America reproduces parthenogenetically[5] and also defends itself against flatworms.[6]

Chapter 9 // Invertebrates Exclusive of Insects

10

/ REPTILES, AMPHIBIANS, AND FISH /

Male bullfrog (*Lithobates catesbeianus*), in an inlet of the Schuylkill River, downtown Philadelphia.

The big target-shaped disk behind the eye of male bullfrogs is a tympanum (hearing organ). In studies, bullfrogs adjusted the timing of their mating calls to avoid interference from noise produced by vehicular traffic.[1] Courtship and mating in the city depend on sensory systems that must cope with urban noise. This chapter presents a wide range of environmental and social stimuli that shape reproduction in urban reptiles, amphibians, and fish.

Anti-aphrodisiac Pheromone: common garter snake (*Thamnophis sirtalis*)

Attempts of male garter snakes to prevent their mates from mating with other males often fail.

A female garter snake produces a sex pheromone that attracts males.[1] They track her by following her pheromone trail.[2] Before copulation, males lick and analyze sex pheromone coating her back.[3] After copulation, she stores her mate's sperm. Storage may last for months before fertilization.[4] During this interval, she may mate again. Sperm from each mate competes for fertilization. Each mate may end up siring only part of her litter. This sperm competition is reminiscent of that noted in nightcrawler earthworms and in cabbage white butterflies.

Her mate deploys several weapons to deter her from mating with potential rivals. His equivalent of a penis (a hemipenis) has recurved spines that hook on to her during copulation. They may delay her from disengaging.[5] He obstructs her genital tract with a mating plug that acts as a "chastity belt"—a temporary barrier to copulation.[6] He impregnates her not only with sperm, but also with anti-aphrodisiac pheromone, which temporarily reduces her attractiveness to other males. Males can detect this pheromone at a distance.[7] As noted earlier, male bedbugs, cabbage whites, and bumblebees also produce anti-aphrodisiac pheromone.

Photo by Frank Miles

Mating aggregation of the common garter snake in spring in an arboretum near Philadelphia. The two smaller snakes are males courting the larger snake, which is female.

Males often fail to prevent female garter snakes from remating. Genetic testing in several studies has revealed multiple paternity in most litters.[8] In cities, production of a litter sired by two or more males may be particularly advantageous. Urbanization may genetically isolate populations of garter snakes (as in toads). Common garter snakes avoid crossing roads. Pavement degrades their ability to track sex pheromone.[9] Visitors in municipal parks undermine genetic diversity of garter snakes by collecting them.[10] Genetic isolation has caused urban populations of garter snakes to inbreed and to lose genetic diversity.[11] Polyandry in common garter snakes theoretically strengthens genetic mixing.

Mimetic Sex Pheromone: common garter snake (*Thamnophis sirtalis*)

Female garter snakes attract male garter snakes, but some male garter snakes attract male garter snakes.

Female garter snakes emerging from hibernation in spring attract courting males. Males aggregate around them. By contrast, some males aggregate around males that mimic females. These male mimics produce a sex pheromone that attracts males. They do not court females, or they do so only sluggishly. They are slow, and they are vulnerable to predators because they do not flee danger.[1]

Common garter snake. The nostril is conspicuous, but the male detects attractive sex pheromone not through his nose, but with his tongue and a sensory organ on the roof of his mouth.[5]

Male garter snakes emerge from hibernation before females. They begin seeking females after they have acclimated, which takes a day or two. Male mimics of females constitute a transient physiological state during this period of acclimatization. Richard Shine, an evolutionary biologist at the University of Sydney, and colleagues hypothesized that aggregation of males around these mimics transfers heat to the mimics. He found that experimental elevation of body temperature of mimics by 13 degrees Celsius prevented them from producing pheromone attractive to males, and it transformed them into functional males. Unlike mimics, these transformed males vigorously court females, move quicker, and flee predators. Shine's team found that experimentally lowering body temperatures of these males restored them to mimics of females.[2]

Male garter snakes attracted to a mimic may serve the mimic in other ways. The investigators noted that a mimic is lethargic and vulnerable to attack by crows preying on garter snakes emerging from their winter dens. They hypothesized that acclimated fellow males protect the mimic.[3]

Male protection of fellow males transitioning from hibernation may be widespread in garter snakes. In Missouri, male garter snakes that had recently emerged from hibernation collectively formed a guard at the entrance to their den. Six of them were photographed close together with their heads elevated high off the ground. Each sentinel in this tight formation faced out in a different direction.[4]

Sex Determination by Temperature: red-eared slider turtle (*Trachemys scripta elegans*)

Temperature of eggs during self-incubation determines sex of developing red-eared sliders.

Egg-laying female.

Juvenile.

Hatchling, still covered with dirt.

The red-eared slider is a semiaquatic freshwater turtle native to the midwestern United States and Mexico. It has been introduced throughout the United States and in seventy-six other countries.[1] Commercial trade in pets is responsible for its global dispersal.[2] In urban locations, it is most conspicuous basking on logs or rocks in water. When approached, it is quick to slide into the water in accordance with its common name, "slider."

The female uses her hind feet to dig a shallow hole in an open, unshaded area that is not muddy. This nest site may be close to water or as far away as 1.6 kilometers (1 mile). She deposits her eggs in the hole, which she plugs with dirt. She leaves her eggs to incubate on their own, without maternal care. The eggs hatch in 68 to 70 days. Hatchlings find their way to water.[3]

The red-eared slider has no sex chromosomes.[4] Temperature during self-incubation of eggs in the nest determines whether the embryo will develop as male or female. In laboratory experiments where temperature was kept uniform over time, incubation of eggs of red-eared sliders at a degree below 29 degrees Celsius (84 degrees Fahrenheit) produced exclusively males, and at a degree above, exclusively females. Incubation at exactly 29 degrees Celsius produced both sexes.[5]

The red-eared slider is abundant along the Schuylkill River in Philadelphia. The presence of hatchlings and juveniles demonstrate that, at least in the short term, the red-eared slider continues to reproduce in the city despite urban and global warming.

Temperature-dependent sex determination occurs in some fish and in at least some species in every order of reptiles except snakes.[6]

Amplexus: American toad (*Anaxyrus americanus*)

Only a small minority of males congregating around a pond succeed in mating.

On the first warm nights of spring, American toads migrate to ponds where males can be seen mounted on females in a position called amplexus, shown in the photo. An individual toad tends to return annually to the same pond. In studies, toads oriented toward a recording of a toad chorus.[1] Their high-pitched trilling pierced through low-pitched noise generated by vehicular traffic. Traffic noise did not affect their singing, but it masked lower-pitched calls of green frogs (*Rana clamitans*), which responded by changing the pitch and timing of their calls.[2]

Amplexus. Male is on top. Photographed in a suburban garden pond.

When concentrated around ponds, male toads attempt to mount any suitably sized object that moves, including other males. The male American toad demonstrates no capacity for distinguishing male from female until after he has mounted and embraced a male, at which time a vibratory signal or chirp from the other male induces the embracing male to release his grip.[3] Male American toads freely embrace American bullfrogs (*Lithobates catesbeianus*), pickerel frogs (*Lithobates palustris*), and green frogs.[4]

Rival males try to dislodge males mounted on females. A mating pair defends itself by kicking away the second male. In one study, interlopers succeeded less than 2 percent of the time.[5] Rivals may mount on top of mating pairs and produce trios, with females on the bottom. More males may pile on, producing a "mating ball" that sometimes kills the female.[6]

In studies, females solicited up to half of pairings; they approached and touched males.[7] Females favored males that were large or that sang more often and longer.[8] They recognized siblings according to their song and avoided mating with them.[9] During amplexus, males emitted a distinctive series of clicks that synchronized spawning with sperm emission. Fertilization of eggs took place in the water.[10] Females left the water after spawning, but males lingered, ready to mate again.[11]

Early departure of females from ponds shifts the sex ratio. In studies, only about 10 percent of toads around a pond were female.[12] Rates of fertilization of eggs approached 100 percent, but the fraction of males that succeeded in mating around a pond was only about 20 percent.[13]

Alternative Mating Tactics: American toad (*Anaxyrus americanus*)

Small male toads mount females before, rather than after, the females reach water.

To improve their chances of reproduction, male American toads try alternative tactics for mating. They position themselves outside ponds and intercept and mount incoming females on land. Once mounted on a female, the male holds on as she completes her journey to the pond. Males that adopt this "gauntlet" strategy are on average smaller than males in the pond. In one study, males intercepted 70 percent of females before they reached a pond.[1]

Trying yet another tactic, a small male may maneuver himself in water to fertilize eggs of a spawning pair in amplexus. Ordinarily, a male in amplexus releases his sperm while his mate extrudes her eggs into the water. He cups his hind webbed feet to guide his sperm onto her eggs. A small, poorly competitive male may exploit a vulnerability in amplexus. He may back up in the water to the mating pair and align his cloaca (rear end) against the female's. As the mating pair moves, he moves with them, maintaining cloacal contact. He sings repeatedly during these maneuvers. This cuckolder may succeed in fertilizing eggs of a female that he has not mounted.[2]

Female approaching a suburban pond to mate. A male may intercept her before she reaches it.

Susan Hitchings and Trevor Beebee, molecular ecologists at the University of Sussex, described genetic losses in small urban populations of the common toad (*Bufo bufo*) isolated by roads and development in Brighton, England. Genetic diversity and fitness were low in these populations compared to nearby rural populations. Hitchings and Beebee anticipated that continued loss of genes during sexual reproduction would cause these small, isolated urban populations to go extinct.[3]

Other species with alternative tactics for mating described in this book include the eastern gray squirrel, European paper wasp, and bluegill.

Panmixia: American eel (*Anguilla rostrata*)

The global population of American eels migrates to mix and spawn in the Sargasso Sea.

The American eel matures in rivers, lakes, and streams, and in coastal marine and brackish waters ranging from Venezuela to Labrador and Greenland.[1] Maturation may take two or three decades or longer.[2] Mature American eels migrate to the Sargasso Sea (near Bermuda) to mix, spawn, and die. Genetic mixing encompasses the global population as a whole rather than geographic subpopulations. Eggs and larvae disperse back in ocean currents toward coasts of North America.[3]

Nobody has witnessed spawning of the American eel in the wild. A century ago, the Danish biologist Ernst Johannes Schmidt deduced the location of its spawning grounds based on me-

Photo credit: Leo Sheng

American eel caught and released in Schuylkill River, Philadelphia.

thodical ocean sampling of the young of American and European eels (*Anguilla anguilla*).[4] This century, Mélanie Béguer-Pon at Université Laval (Quebec) and Dalhousie University (Halifax), and colleagues, tracked migrating American eels fitted with satellite tags. One of the eels they tracked traveled 2,400 kilometers in 45 days from Nova Scotia to the northern limit of its presumptive spawning area in the Sargasso Sea.[5] Molecular genetics of young eels from locations distant from one another demonstrated panmixia—random genetic mixing of all reproductive members of the species.[6]

How do American eels accomplish the navigational feat of finding one another in the Sargasso Sea, in some cases thousands of kilometers from their point of departure? They may integrate many cues: geomorphic, geomagnetic, thermal, circadian, photoperiodic, lunar, chemical (saline concentration), hydrographic (currents), pheromonal, olfactory, and visual.[7] They may spawn in large aggregations, amplifying broadcast of their individual visual and olfactory signals.[8]

Steven Shepard of the U.S. Fish and Wildlife Service noted that habitats of the American eel are more diverse than those of any other North American fish.[9] Panmixia suggests why. This mixes genes for adaptation to places ranging from Greenland to Guyana. It theoretically prepares the species for diverse and unpredictable conditions, including those found in cities. But panmixia also makes the American eel vulnerable. Dams and weakening of the Gulf Stream threaten the American eel's migratory life cycle. Numerous disturbances threaten its spawning grounds in the Sargasso Sea.[10]

Parenting Alone: male bluegill (*Lepomis macrochirus*)

Nesting male bluegills do all the work of building nests, attracting mates, nurturing broods, and protecting fry.

Bluegills are among fish reeled in from the Schuylkill River in downtown Philadelphia during a "fishing fest" to promote public appreciation of clean water.

In spring, the male bluegill constructs his nest alone. Stabilizing himself with his pectoral fins, he moves his tailfin back and forth like a broom, sweeping out a saucer-shaped depression in sand or gravel. Typically, he excavates his nest near other nests, which together form a colony resembling a cluster of craters.[2] After completing his nest, he swims around its rim in circles, signaling to egg-laden females hovering nearby in deeper water.[2]

When a female bluegill swims into his nest, he encircles her, and she may join him circling around the rim of the nest. Sometimes the female swims away and he chases her. He may encircle her and guide her back into his nest. Once she is settled into his nest, he swims beside her, facing the same direction. Inside the nest, they circle together, he on the outside and she on the inside. She tilts her top away from him and her bottom toward him. She vibrates against him near his tailfin and deposits eggs, and presumably, he synchronizes his discharge of sperm.[3]

After spawning, his mate leaves. He defends the nest against predators. He nips the fins or gill covers of rivals and chases them away. He fans his brood with his fins, maintaining circulation of oxygenated water.[4]

He is sometimes able to detect by scent that a cuckolder has sired part of his brood.[5] He provides parental care preferentially to his own fry.[6] He may cannibalize fry sired by others.[7] Based on a paternity survey in one colony, nesting males sired from 26 to 100 percent of fry in their nests.[8]

Sexual Mimicry: male bluegill (*Lepomis macrochirus*)

Some male bluegills cuckold other males by mimicking females.

Some male bluegills do not make nests or provide parental care. They are specialized cuckolders. Compared to nesting males, they reach sexual maturity sooner and remain smaller.[1] Cuckoldry in these males is not a transient developmental stage but rather lifelong. Cuckolders in bluegills come in two forms: "sneakers" and "satellites." Sneakers mature into satellites, but neither matures into nesting males. In one population, sneakers peaked in abundance at age two, satellites at age four, and nesting males at age eight.[2]

Satellites mimic females in coloration and behavior. In one study they measured on average 3 centimeters smaller than females and 6.4 centimeters smaller than nesting males. They hover above a nesting male and descend slowly into the nest while a female is present. A satellite may interpose himself between the female and nesting male and discharge his sperm as she releases her eggs. When the nesting male acts aggressively toward the mimic, the mimic exaggerates his female-like behavior or flees, repositioning himself as a satellite above the nesting male.[3]

Sneakers use a different tactic. They are even smaller than satellites. They hide behind rocks and plants just outside nests and wait for an opportune moment when a female is in the nest and the male is distracted defending his nest. The sneaker will then dart into the nest, position himself beneath the spawning female, pause momentarily, release his sperm, and flee—all within 10 seconds. Several sneakers may intrude simultaneously or in rapid succession, diverting the nesting male.[4]

These cuckolders distinguish themselves by their extreme physical specialization. They exploit nesting males but support their species by increasing genetic mixing. Theoretically, this support may apply especially in cities. Anglers may toss them back in the water due to their small size. Due to precocious maturation, they can quickly increase in number and service females after a population has crashed. They attend to spawning females when nesting males are distracted. Experimentally, when exposed to a predator, their offspring survived better than nesting males' offspring.[5]

Bluegill, labeled as female. Some male bluegills mimic the behavior and appearance of females. (Chromolithograph from *Fishes of North Carolina*.[6])

Feminizing Pollution: bluegill (*Lepomis macrochirus*)

Pollutants from diverse sources drain into rivers and disturb sexual differentiation in fish.

Some pollutants in rivers behave as estrogens, feminizing male fish. They can cause intersexes in which male gonads contain oocytes—cells ordinarily confined to ovaries. Oocytes are capable of forming eggs.

A large survey on sexual effects of pollution on fish was undertaken in rivers in North Carolina. Ten percent of bluegills surveyed were intersexes. The prevalence of intersexes correlated with the level of exposure to endocrine-active compounds suspended in the water column. These compounds included estrogens, organochlorine pesticides, metolachlor, and triazines. Feminization was magnified in bass (*Micropterus* spp.). Unlike bluegills, bass prey on fish and thus concentrate pollutants up the food chain. The prevalence of intersexes in bass was 60 percent; it correlated with the amount of pollutants deposited in the riverbed.[1]

An open question is whether feminization of bluegills due to pollution reduces their reproduction.

Drain to sewage system that combines storm water and household sewage into one pipe. In major storms, household sewage may overflow into the river.

11

/ MAMMALS /

House mouse (*Mus musculus*) feeding on seeds that have spilled from a bird feeder.

Courtship sounds emitted by house mice are ultrasonic.[1] The eastern cottontail rabbit is sexually silent except for squeals during copulation.[2] By contrast, both male and female eastern gray squirrels produce mating calls audible to human beings.[3] In these animals, diversity of courtship contrasts with similarities in lactation and nursing. This chapter examines a particular sexual reproductive trait that helps to sustain urban populations of the eastern cottontail. It explores pheromones and sexual behavior that support genetic diversity in the house mouse. By contrast, it considers a sexual trait that has contributed to mortality in an urban population of the eastern gray squirrel.

Territoriality: eastern gray squirrel (*Sciurus carolinensis*)
The eastern gray squirrel is not territorial.

Squirrel nest in a rowhouse garden downtown.

Male animals commonly express sexual dominance through territoriality, as in the male blue dasher dragonfly. The eastern gray squirrel is an exception. Neither the male nor the female eastern gray squirrel defends territories. Within their home ranges, males adhere to hierarchies of dominance, especially at feeders; but home ranges overlap: two or more males may occupy the same parcel of land.[1]

Home ranges of females likewise overlap. Females defend core areas immediately around their nests, but not territory such as space dedicated to foraging. They protect their nests from females unrelated to them, and from males (although males sometimes help them build nests).[2] Related females nest communally; unrelated females nest together only rarely. Males nest together or alone, usually separate from females.[3]

Lack of territoriality enables the eastern gray squirrel to achieve high population densities. Densities can increase to match availability of food. However, in urban parks the result can subject squirrels to crowding. Lafayette Square is a 3-hectare (7-acre) public park across the street from the White House in Washington, D.C. From 1980 to 1985 the squirrel population in this park frequently exceeded 100, causing mortality from fighting, wounds, and infectious disease. Park managers discovered that visitors had been feeding the squirrels over 34 kilograms (75 pounds) of peanuts each week. The managers eventually intervened.[4]

Chapter 11 // Mammals

Mating Plug: female eastern gray squirrel (*Sciurus carolinensis*)

The male eastern gray squirrel inserts a mating plug into the genital tract of his mate.

The mating plug has been interpreted as functioning like a chastity belt: a physical barrier blocking insemination by other males, at least temporarily. It may also function as a "nuptial gift," transferring nutrients or other beneficial substances to mates. It may prevent leakage of sperm; or alternatively, it may control release of sperm. It may deliver hormonally active compounds, or pheromones, such as anti-aphrodisiac pheromone. Among animals presented in this book, those using mating plugs include cabbage whites, fireflies, bumblebees, and garter snakes.[1]

The mating plug in the eastern gray squirrel is a waxy or rubbery object apparently formed from seminal fluid. In one study, females removed mating plugs half the time within 30 seconds of copulation. They usually ate the plugs, but sometimes they discarded them.[2] Mating of a female with multiple males serves her interests by helping to ensure her fertilization.[3] By contrast, polyandry opposes her mate's interests by reducing his chances of paternity. The mating plug is a source of sexual conflict.

Female eastern gray squirrel,

Alternative Mating Tactic: male eastern gray squirrel (*Sciurus carolinensis*)

Subordinate males of the eastern gray squirrel resort to "scramble competition" for mates.

Male eastern gray squirrel.

In mating chases, one or more dominant male eastern gray squirrels run at high speed in pursuit of a female in estrus. Males in the mating chase are older compared to males on the sidelines. Copulation takes place in trees, often high in the tree canopy, after a male leading the chase corners a female at the end of a branch. Competing males sometimes knock them off the tree.

Fewer than a third of males in a population engage in mating chases. The remaining males wait opportunistically nearby. Designated satellite males, they are younger, smaller, and socially subordinate. They mate with females who bolt from mating chases. While fleeing, such females mate with the first male they encounter. Satellite males scramble to seize the opportunity. Males from aborted mating chases sometimes join this "scramble competition."

What induces females to pick subordinate over dominant males? John Koprowski at the University of Kansas investigated these chases. He hypothesized that females break away from dominant males to escape danger. The mating chase, with its high-speed arboreal leaps and turns, puts the female at risk of injury and even death.[1] Her selection of a less aggressive mate may also select for offspring who are less aggressive. As noted earlier, crowding of squirrels in Lafayette Park in Washington, D.C., caused fighting and increased mortality. An open question is whether crowding in urban settings generates selective pressure for evolution of loss of aggressiveness, reminiscent of loss of aggressiveness in urban house finches (page 104) and bridge spiders (page 78).

Copulation-Induced Ovulation: eastern cottontail (*Sylvilagus floridanus*)

Female eastern cottontails ovulate 10 to 11 hours after copulation.

The eastern cottontail nests on the ground in cavities at a depth of less than 12 centimeters (5 inches), where predators, especially dogs, coyotes, and raccoons, attack it. Without powerful jaws and claws, the eastern cottontail's best defense is hiding or fleeing.[1] Unlike squirrels, it cannot escape by climbing trees; and unlike the European rabbit (*Oryctolagus cuniculus*), it does not dig burrows.[2] Numerous parasites and pathogens attack it. In a city park in Chicago, annual survivorship of eastern cottontails averaged just under a third.[3]

The eastern cottontail's chief defense against high mortality is high fecundity. In a Missouri wildlife conservation area, females each produced in the course of one breeding season an average of 35 young, distributed across seven to eight litters.[4] From April through June, almost all females in this area were pregnant.[5] Because female rabbits store sperm only briefly, they must closely synchronize ovulation with insemination. Ovulation here refers to release of eggs from ovaries into uterine (fallopian) tubes.

The eastern cottontail ovulates 10 to 11 hours after copulation.[6] Sperm penetrate eggs two to three hours after ovulation (as assessed in the European rabbit).[7] The mechanism by which copulation in rabbits induces ovulation has been elusive. Behavioral sexual stimulation of female rabbits without copulation has induced ovulation.[8] When injected, a preparation made from rabbit semen induced ovulation—but only in llamas.[9] Whatever may be the mechanism, copulation-induced ovulation in rabbits supports fecundity and helps offset high urban mortality.

Camouflaged eastern cottontail along a bicycle path in downtown Philadelphia.

Pregnancy Blocker: house mouse (*Mus musculus*)

A male sex pheromone of the house mouse blocks pregnancy.

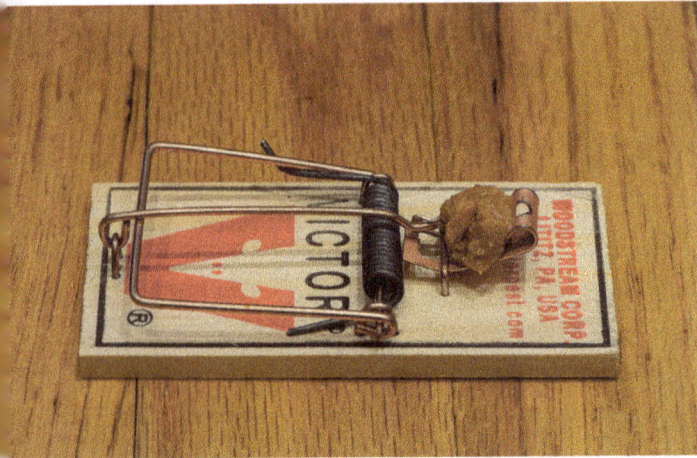

A potential disruptor of a dominant male or a nursing mother house mouse.

The house mouse lives indoors or outdoors in groups of two to a dozen adults. The dominant male in a group attacks potential rivals, sometimes mortally, and he chases away young as they mature. He defends the group's territory. In dense colonies several mothers and their litters may nest communally. The dominant male sires the majority of litters.[1]

Wesley Whitten at Australian University noted that a male house mouse experimentally introduced into the cage of a female did not mate on the first night but rather on the third. He determined that presence of a male induces estrus in the female. He showed that air can transmit to the female the male pheromone that induces her estrus.[2]

Soon after Whitten's discovery, Hilda Bruce at the National Institute for Medical Research in London discovered a male sex pheromone that blocks pregnancy. She found that exposure of a newly mated female mouse to a strange male mouse blocked her pregnancy and allowed her to remate in three to six days. Mere placement of a newly mated female in a cage soiled by a male of a different genetic strain blocked pregnancy.[3] The pregnancy-blocking effect of this male pheromone serves the strange male's interests by freeing up a potential mate. It serves the female's interests by freeing her to mate with a male that is genetically novel. In confined quarters such as urban houses, one might expect that house mice would be vulnerable to inbreeding. This pregnancy blocking effect promotes outbreeding. It also protects against infanticide. A strange male intruder may kill young sired by others.[4]

Female house mice that nest communally may also nurse communally. Communal nursing is potentially most advantageous to mice that are unlikely to succeed in nursing on their own. Communal nursing requires synchronization of estrus.[5] Sex pheromones regulating timing of estrus provide a means for this synchronization.[6] Conceivably, communal nursing buffers a litter against death or illness of one nursing mother.

In a paternity survey of litters of wild house mice, more than one male sired a litter in about a quarter of cases.[7] Failure of dominant resident males to monopolize females in their nests promoted genetic diversity. This behavior and pregnancy-blocking pheromone both support genetic mixing. The breeding system of the house mouse in the city is well adapted to coping with isolation of populations in buildings.

12

/ BIRDS /

Courtship feeding by the northern cardinal (*Cardinalis cardinalis*). The male on the left is feeding a female, probably his mate.

The pages that follow illustrate courtship feeding also in mourning doves and house finches. While many species engage in courtship feeding, reproductive behavior in birds in the city is nevertheless diverse. A male Canada goose attempting to copulate with a female Canada goose outside social pair bonds may incite a violent struggle. A repeatedly widowed red-tailed hawk in the city bonds with a succession of male suitors. The female brown-headed cowbird induces other species of birds to raise her young. The white-throated sparrow mates as if it has four sexes. This chapter delves into each of these examples.

Territoriality: red-tailed hawk (*Buteo jamaicensis*)

Territoriality in the red-tailed hawk varies with circumstances.

Family of red-tailed hawks nesting on a customized ledge on the façade of the Franklin Institute in downtown Philadelphia, May 26, 2012. The male in this photo had recently replaced the nestlings' biological father, who was killed in a highway accident.

In contrast to the eastern gray squirrel, the red-tailed hawk is fiercely territorial. And unlike the female blue dasher dragonfly, the female red-tailed hawk may join her mate in territorial defense.[1] Defense may involve only females, only males, or both.[2] The red-tailed hawk monitors its territory from a lookout, such as a perch on a branch of a tall tree, or watches while soaring.[3] A female may breed in the same territory over a succession of years.[4]

In 2009, a pair of red-tailed hawks nested on a ledge of the façade of the Franklin Institute in downtown Philadelphia. They continued to raise broods at the site until the spring of 2012, when the male died in a highway accident. The female accepted a second male, who helped raise the brood sired by the first, as shown in the photo. After two years, she took on a third male after the second died in a railroad accident. They nested in an abandoned squirrel nest in a tree nearby. Two years later, she accepted a fourth mate after the third died from rat poison.[5]

Pair bonds of red-tailed hawks are thought to be lifelong, but death encourages formation of new pair bonds.[6] Flexible territoriality makes it possible for an urban widow to bond with an intruder.

Conditional Sexual Trait: house finch (*Haemorhous mexicanus*)

Sexual coloration in male house finches depends in part on diet.

In experiments, female house finches favored males colored brightest red. In the field, wild males that had paired with females were brighter than males that had not paired.[1] The brightest males paired earlier in the breeding season than did other males; as a result, they produced more offspring in the course of a year.[2] Males artificially brightened with red dye paired sooner than did males treated with dye that lightened or matched their color.[3]

When wild male house finches from different geographic areas were caught and maintained in captivity on a standardized diet, differences in their redness almost disappeared. Elimination of dietary carotenoids (plant pigments) caused their feathers to turn yellow.[4] Experimentally, house finches given dietary supplements of carotenoids mounted stronger immune responses against bacteria inoculated into their blood.[5]

Male.

Bright red coloration in plumage of the male house finch was associated with resistance to bacterial conjunctivitis, which in house finches has caused lethal epidemics.[6] In male house finches, red coloration correlated positively with fecundity, winter survival, and paternal care, and negatively with abandonment of nests by female nest mates.[7] Brighter males had higher levels of testosterone.[8]

Female.

According to Darwin's idea of sexual selection, female preference, reinforced over countless generations, drives evolution of bright male colors. The house finch at first seems to illustrate this idea. Darwin, however, presented female taste for male ornamentation as purely esthetic, like the taste of a poultry fancier.[9] By contrast, the house finch shows how female preference of sexual coloration is utilitarian, selecting for male fitness.

On the other hand, female house finch preference for bright red in mates persists in populations where male sexual coloration is muted.[10] Her preference may linger even after red fades as a marker of male quality. Her taste for red raises the possibility that a sense of sexual beauty that starts off utilitarian may in time become purely esthetic, as Darwin imagined.[11]

Submissive Male: house finch (*Haemorhous mexicanus*)

In urban house finches, colorful males are less aggressive than dull males.

Male house finches fighting at a bird feeder.

House finches are gregarious and semicolonial. They forage in groups and socialize in a pecking order, with subordinates avoiding those ranked higher. Aggressive behavior includes attack and displacement, "beak fencing," and vigorous struggle.[1]

In studies of competition for food and perches among caged house finches, colorful males were paradoxically less aggressive than dull males.[2] Masaru Hasegawa and colleagues, behavioral ecologists at the University of Tsukuba, Japan, and Arizona State University, found that low aggressiveness in colorful males applied only to urban males. Colorful rural males were not less aggressive than dull rural males. Among colorful males, aggressiveness was less in urban than rural males. Among dull males, aggressiveness in urban and rural males was the same.[3]

Other studies found that bills of urban house finches are bigger than those of nearby rural house finches. Bills of urban birds have evolved to crack sunflower seeds at bird feeders. Rural bills have evolved to handle smaller seeds, which predominate in the wild.[4]

Perhaps low aggressiveness, like bigger bills, is an evolutionary adaptation to urban bird feeders. Low aggressiveness may protect against injury during fighting at feeders. In the city, colorful males compared to dull males may be not only better nourished and healthier, and better mates, but also better socialized. Hasegawa's team considered this hypothesis and others to explain submissive behavior of colorful urban males.[5]

Courtship Feeding: house finch (*Haemorhous mexicanus*)

Male house finches feed females.

When the unpaired male red-tailed hawk shown in the photo on page 102 first visited the nest of his newly widowed prospective mate, he presented her with an offering: a mouse. Courtship feeding in birds is widespread, particularly in socially monogamous species in which both sexes feed the young. The term has been applied to any instance of an adult male feeding an adult female; such instances may occur after courtship, such as on the nest during incubation.[1] Courtship feeding may also present as one of a series of stereotypic rituals immediately preceding copulation, as shown in the sequence of photos of mourning doves on pages 108 to 110.

The British evolutionary biologist and ornithologist David Lack concluded that the primary function of courtship feeding was not nutrition. He cited a female British robin (*Erithacus rubecula*) begging her mate for food as she stood on her feeding tray filled with mealworms. He interpreted courtship feeding as strengthening the pair bond.[2] Other researchers viewed it as supporting high-energy tasks. They noted that courtship feeding has evolved most often in noncarnivorous species in which the female alone builds the nest and incubates.[3]

The house finch has all the traits of a typical courtship feeder.[4] However, the functional significance of courtship feeding shown in the photo is ambiguous. Not shown is a nearby feeder filled with shelled sunflower seeds, a favorite of the house finch. Perhaps benefits of courtship feeding shown in the photo are both social *and* nutritional.

Courtship feeding.

Repeated Broods: mourning dove (*Zenaida macroura*)

A pair of mourning doves may attempt to make as many as six broods per year.

Female.

Male.

Life expectancy of adult mourning doves varies according to region but may average only about a year.[1] In Berkeley, California, the annual mortality rate of mourning doves residing in the city was almost 50 percent—similar to that in mourning doves subjected to hunting.[2] On the campus of Texas A&M University, only about a quarter of nests of mourning doves produced fledglings.[3]

To maintain its population despite high mortality and brood failure, the mourning dove attempts to breed often. It compresses its brood cycle to 28 to 30 days, defined as the period from laying the first egg to fledging of the last young. It nests up to six times annually. It begins a new clutch within six days of completion of nesting. Sometimes it begins a new clutch before its current clutch has fledged.[4]

Members of the dove family (Columbidae) are remarkable for producing multiple broods in rapid succession. The small size of their broods helps these birds cycle through quickly.[5] The clutch of a mourning dove contains only one or two small eggs.[6]

To space broods close together, doves must avoid delays tied to conception. Mei-Fang Cheng at Rutgers University found that female ring doves began repetitive copulation before their period of fertility. They continued to copulate during their period of fertility.[7] Theoretically, repeated copulation starting before ovulation advances time of conception to the earliest possible moment. It may help doves maximize the number of broods they can squeeze into their breeding season.

Repetitive copulation in birds is common and has inspired many explanatory hypotheses, which may all be true.[8] In one study, northern goshawks (*Accipiter gentilis*) copulated on average more than 500 times per clutch. They started copulating almost two months before laying their eggs.[9]

Shortened Courtship: mourning dove (*Zenaida macroura*)

A pair of mourning doves shortens courtship after it forms a pair bond.

Pair bonds formed by mourning doves last one season or longer. This duration spares the birds the task of finding new mates and establishing fresh pair bonds before each brood. For experimentally caged mourning doves, courtship to establish initial pair bonds took 11 days.[1] The sequence of photos on pages 108 to 110 shows courtship of two mourning doves in our garden on April 30, 2017. By this date, the two had likely completed their initial courtship, which may have included charging, bowing, and leaping.[2] At the time the photos were taken, they performed none of these rituals. The time between the first photo and copulation was under four minutes. Short-

A pair of mourning doves moments before initiating courtship. The female is on the left.

ened courtship allowed the pair to attempt more copulations before each brood. As noted on the previous page, this strategy could shorten their brood cycle, increasing their number broods.[3]

The photos illustrate stereotypic courtship behavior:

+ Preening (figures 3 and 4): Each member of the pair preens while the other watches.

+ Male display (figures 5, 6, and 7): He puffs himself up and ruffles his feathers as she observes.

+ Courtship feeding (figures 8 and 9): She crouches and sticks her bill inside his. This form of courtship feeding resembles the feeding of young, when a mourning dove regurgitates "crop milk" that nestlings imbibe from inside its beak.

+ Billing (figure 10): The sides of the beaks of the two sexes touch.

+ Allopreening (figure 11): Billing transitions directly into allopreening. Here, the male applies his bill to his mate's neck. Allopreening in courtship is often the last ritual before copulation, as it is here.

+ Postcoital indolence (figures 13 and 14). In the aftermath of copulation, the pair rests close together.

Fig. 1

Fig. 2

Fig. 3

Fig. 4

Fig. 5

Fig. 6

Fig. 7

Fig. 8

Fig. 9

Fig. 10

Fig. 11

Fig. 12

Fig. 13

Fig. 14

Coping with Noise: mourning dove (*Zenaida macroura*)

Mourning doves avoid nesting near noise that masks their coos.

Like light pollution, noise pollution is characteristic of cities. Clinton Francis at the University of Colorado and colleagues investigated effects of noise on nesting and abundance of birds. A compressor in their study generated the noise. They discovered that noise repelled mourning doves, whose low-pitched coos the noise masked. By contrast, noise attracted small songbirds like house finches, whose high-pitched songs rose above the frequency of the noise. The noise repelled the house finch's primary nest predator, the western scrub jay. By providing protection from an enemy, noise offered the house finch a safe haven.[1] This is reminiscent of the theoretical safe haven that light pollution afforded the common eastern firefly (page 66).

Interstate 76, generator of low-pitched noise.

Male mourning dove in our garden.

Noise may affect not only abundance and nesting of birds, but also their songs. In response to noise, both doves and house finches alter their songs in a way that reduces masking.[2] Their response to noise is similar to that of frogs (pages 85 and 89).

Noise may interfere with sexual physiology. Mei-Fang Cheng at Rutgers University and colleagues studied coos in the ring dove. They found that coos generated neuronal output to the hypothalamus, a region of the brain with hormonal links to the pituitary gland. They measured sex hormone (luteinizing hormone) in veins draining the pituitary gland. Coos specific to the ring dove increased release of this hormone by the pituitary gland. White noise blocked this increase.[3]

Noise from traffic has reduced clutch size and number of fledglings in birds.[4] Traffic noise has masked acoustic signals of stream fish.[5] In aquarium experiments, noise has impeded spawning of fish.[6] Traffic noise has interfered with courtship signals of grasshoppers and crickets.[7] Vibratory noise has interfered with predatory behavior of spiders; effects of vibratory noise on reproduction in spiders has yet to be studied.[8] An open question is how noise produced by traffic and other urban sources affects urban ecosystems as a whole.[9]

Sexual Singing: American robin (*Turdus migratorius*)

The mated male American robin sings when his mate is on the nest.

Adult.

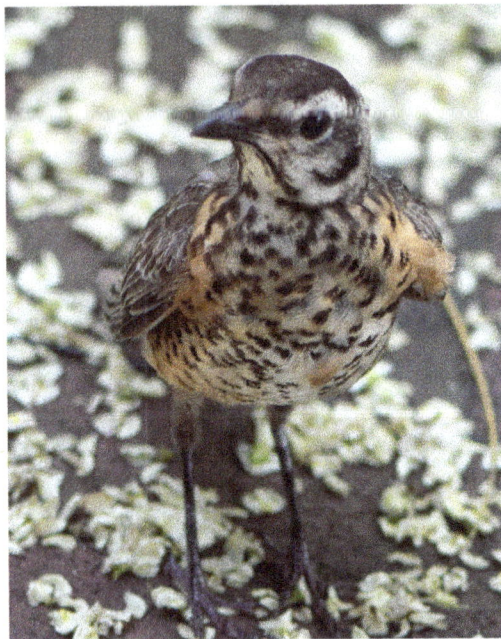

Fledgling.

Among American songbirds, the American robin is often the first to sing in the morning and the last in the evening.[1] In cities, it adjusts its singing according to noise and light pollution.[2] Spectrograms have described its songs and their many variants.[3] Ornithologists have long presumed that the function of dawn singing of the American robin is sexual. Study of sexual behavior of the robin has supported this view.

During the last two weeks of April, Tore Slagsvold, ornithologist from the University of Oslo, daily visited territories of 12 mated male American robins in a park near Seattle. He monitored them before they began to sing before sunrise. He continued his observations until noon, and then again at dusk until darkness.

He found that male singing peaked at dawn just before females left their nests to forage. Singing mostly stopped soon after females departed. To a lesser extent males sang during the day when their mates occasionally visited their nests, and after sunset when their mates returned to their nests to roost for the night.

Slagsvold suggests that dawn singing of the mated male American robin serves one or more functions. It may:

1. Reinforce pair bonding and sex hormonal stimulation.

2. Signal his mate that she can safely leave to forage.

3. Attract his mate for copulation.

4. Attract other females for copulation.

5. Attract a new mate in case the original one disappears.

6. Repel rivals.

7. Support his territoriality.[4]

Regardless of these functions, the American robin does produce offspring in cities.[5] Evidence of this in downtown Philadelphia is the appearance of fledglings.

Four Sexes: white-throated sparrow (*Zonotrichia albicollis*)

A white-throated sparrow picks a mate from only one in four white-throated sparrows.

White-striped form, male or female.

White-striped form, male or female.

Tan-striped form, male or female.

Tan-striped form, male or female.

The term "four sexes" in the white-throated sparrow refers to color, behavior, and genetics rather than anatomy. The white-throated sparrow has four functional sexes: male and female white-striped, and male and female tan-striped. Males of a color form mate with females of the opposite color form. Either sex can be either color form. White-striped males mate with tan-striped females. Tan-striped males mate with white-striped females.[1]

White-striped males act aggressively toward any white-throated sparrow that sings. They chase away singing females—which are always white-striped—and mate with tan-striped females—which never sing. The less aggressive tan-striped male competes poorly against the white-striped male for tan-striped females. White-striped males sing four times more often than do tan-striped males. So the tan-striped males mate with the singing white-striped females, which white-striped males reject. The trill note of the white-striped female evokes copulatory excitation in tan-striped males.[2]

White-throated sparrows overwinter in cities but migrate in spring to breed in boreal forests.[3] In contrast to cities, boreal forests offer this species space to accommodate four sexes, each with a different suite of traits and needs.

Colonial Bonds: House sparrow (*Passer domesticus*)

The house sparrow lives in colonies.

House sparrows. They feed in flocks like house finches.

Male.

Female.

The house sparrow and house finch both forage in flocks. To some extent they compete, but each is distinctive.[1] The British ornithologist James Denis Summers-Smith, a specialist in the study of the house sparrow, found that this bird lives in colonies where it bonds to both its mate and its nest. Should its mate die, it will retain its nest and bond with a new mate who will share the nest. Since nearly half of adult house sparrows die in a year, pair bonding acts at frequent intervals to maintain a colony.[2] Although the house sparrow is socially monogamous, males may nevertheless mate with more than one female, and females with more than one male.[3]

The house sparrow has been declining in abundance in Europe and North America. Populations have crashed and come close to extinction in the centers of London, Edinburgh, and Glasgow. Summers-Smith noted that the causes for these declines are unclear. However, he hypothesized that as colonies shrink and become farther separated from each other, they may stop functioning as breeding units and stop providing the social stimulation that house sparrows need to reproduce. They may also stop making "floaters"—unpaired birds that travel between colonies and bond with widows and widowers. He proposed that decline of a colony beyond a certain threshold triggers its collapse. He pointed out that loss of social stimulation likely contributed to reproductive failure and extinction of the passenger pigeon.[4]

In cities, low population densities may interfere with mating in other species. Examples include the herbaceous annual *Crepis sancta* in the city of Montpellier, France,[5] and possibly yellow toadflax (page 26). Suppression of mating due to a decline in population density is known as an Allee effect, named after Warder Clyde Allee, who described the phenomenon.[6] Population declines due to Allee effects may be self-reinforcing.

Rape: Canada goose (*Branta canadensis*)

Canada geese are less monogamous than they at first appear.

Canada geese bond with mates for life. *Divorce*, defined as the breakup of a bonded pair for reasons other than death, is rare.[1] On the other hand, brood parasitism and extra-pair paternity are both common. *Brood parasitism* occurs when a female Canada goose deposits her egg in the nest of another, who incubates and rears the interloper's chick. In a study of urban Canada geese, brood parasitism occurred in more than one in four broods.[2] Offspring sired by a father outside his social pair bond occurred in one in seven broods. This rate of extra-pair paternity is comparable to that in other species of socially monogamous birds. Among birds generally, extra-pair paternity has been documented in over 90 percent of species tested.[3]

Rape in Canada geese has been defined as copulation of an unpaired gander with a female paired to another gander. A less judgmental term is *forced copulation*. Both terms refer to vigorous resistance of females to copulation. Two cases of attempted forced copulation were described in which the paired gander was distracted defending his territory. The intruding ganders ran up from behind the females, grabbed the back of the base of their necks, pushed their necks downward, and attempted to copulate. The females flapped their wings, attempted to run away, and made distress calls, alerting the paired ganders, who drove off the intruders.[4]

Forced copulation is more common in waterfowl than in perching birds (songbirds).[5] Female perching birds commonly *solicit* copulation outside pair bonds. Margo Adler at the University of New South Wales hypothesized that resistance of a female to forced copulation selects for vigor in mates outside her pair bond.[6] By selecting for male vigor, she serves the paternal genetic interests of her offspring.

Family of urban Canada geese. The biological parents of some of the goslings likely differ.

Domestication of Sex: Canada goose (*Branta canadensis*)

Canada geese historically migrated in spring to breeding grounds in northern latitudes remote from human civilization.

In *American Ornithology*, published in 1814, Alexander Wilson, "Father of American Ornithology," wrote that the Canada goose, although familiar to everyone throughout the country, does not breed in the wild in the United States; it breeds only in unknown sanctuaries far to the north, perhaps beyond the "pole." On the other hand, Wilson observed that hunters had long domesticated wounded Canada geese. Hunters clipped their wings and raised these pinioned (wing-clipped) geese as decoys to lure members of flocks migrating overhead. These domesticated Canada geese became tame. They paired, nested, and reproduced.[1]

Over generations, captive geese lost their cultural connection to their ancestral breeding grounds. Because of philopatry, captive geese home to their birthplace—specifically, where they first learn as goslings to fly.[2] Captivity transferred their reproduction from sanctuaries remote from human civilization to developed property close to home.

In 1935, the use of live decoys for hunting geese in the United States became illegal. Hunters relocated large captive flocks of decoy geese to wildlife refuges.[3] Wildlife managers bred them and dispersed them as goslings.[4] They favored one midwestern subspecies, or variety: the giant Canada goose.[5] Compared to other Canada geese, this goose nested farther south, matured fast, had big broods, achieved high nesting success, had a high survival rate, and migrated only short distances. The variety was docile and well adapted to urban habitats, especially lawns.[6]

Disseminated by wildlife managers, the Canada goose colonized cities.[7] Despite its proliferation in agricultural and urban landscapes, some populations of this species continue to migrate and reproduce far to the north in their ancestral breeding grounds.[8]

Canada geese and mallards in a municipal park (Fairmount Park).

Brood Parasite: brown-headed cowbird (*Molothrus ater*)

Cowbirds drop their eggs into nests of other species, which then raise their young.

Brood parasitism by cowbirds occurs twice as often in urban compared to rural sites, according to one study.[1] A puzzling question is why some species, like the orchard oriole (*Icterus spurius*), almost always accept cowbird eggs, while other species, like the American robin (*Turdus migratorius*), almost always reject cowbird eggs.[2] The American robin removes cowbird eggs by grabbing them in its beak.[3] It distinguishes its own eggs by their larger size, blue color, and lack of spots.[4] Perhaps species that are frequent victims of cowbirds lack defenses against them. Over the past two centuries, the range of the brown-headed cowbird has dramatically expanded east from population centers in prairies west of the Mississippi River.[5]

Cowbirds are not only brood parasites but also nest predators.[6] Two hypotheses attempt to explain this nest predation: (1) According to the "mafia" hypothesis, cowbirds attack nests of birds that eject their eggs; to avoid retribution, their victims have evolved acceptance of parasitism.[7] (2) According to the "farming" hypothesis, cowbirds induce parents of unparasitized nests to nest again. They do this by preying on nests with nestlings in late developmental stages.[8] Cowbird brood parasitism likely triggered evolutionary arms races between the cowbird and its victims.[9] Pressure for such evolution may be intense in cities, where brood parasitism is common.

Cowbird egg ejected onto sidewalk.

Immature cowbird, resembling the adult female..

American robin's nest, not parasitized.

Polygyny: European starling (*Sturnus vulgaris*)

Male European starlings may form pair bonds with more than one female.

A socially monogamous male starling bonds with only one female in a season even though he may copulate with other females and sire their offspring. Tasks he shares with this female include building the nest, incubating eggs, and feeding young.[1] By contrast, a polygynous male starling bonds with two or more females, whom he supports in separate nests. In one study the frequency of polygyny in male starlings ranged from 20 to 60 percent.[2]

Henrik Smith and Maria Sandell at Lund University found that polygyny in the starling is a source of sexual conflict. It serves the male at the expense of his first mate. Polygyny secures for the male more mates and more ways to pass on his genes. It penalizes his first mate by reducing his support and therefore her success in raising her young. For him, the gain outweighs the loss; for her, the loss has no compensatory gain.

Smith and Sandell suggest that the threat of polygyny shapes behavior of the female starling even in monogamy. She defends her mating status by soliciting copulations during egg laying. Copulation induces him to spend time guarding her and diverts him from seeking other mates. She acts aggressively toward other females, even to the point of injuring them. By contrast, he may counter her attempts to block his bonding with another mate. He may ignore her solicitations for copulation. He may chase her away when she interferes with his courting other females. He may position his second nest at a distance to buffer his second mate from his first.[3]

Polygyny has induced female starlings to abandon their nests.[4] For starlings in metropolitan Rennes, France, intense urbanization decreased food that parents delivered to their nestlings, and it depressed weight of nestlings and number of fledglings.[5] Intense urbanization might be expected to increase costs of polygyny, but whether it discourages males from attempting polygyny is unknown.

Nonbreeding plumage of European starling, photographed on December 13th.

13

/ FUNGI, BACTERIA, AND LICHENS /

Lichens discolored this tombstone in the graveyard of St. Peters Church, Society Hill, Philadelphia.

Growth in the outer margins of lichens produced the rings on this tombstone.[1] Lichens are symbiotic partnerships between fungi and photosynthetic organisms such as algae or cyanobacteria. In contrast to a fungus living symbiotically in a lichen, some bacteria live symbiotically inside cells of multicellular organisms. These bacteria are endosymbionts. This chapter explores sex in lichen symbionts and in bacterial endosymbionts. It also examines sex in microbial biofilms and mushroom-forming fungi.

Microbial Mating: bacteria

Bacteria reproduce without sex, and they have sex without reproduction.

Sex defined broadly applies not only to plants, animals, and fungi but also to bacteria. Jianping Xu, molecular ecologist at McMaster University, defined sex as "any natural process that combines genes from more than a single source in an individual cell."[1] According to his definition, sex may take place in any cell, including the cell of a bacterium. It may occur independent of reproduction.

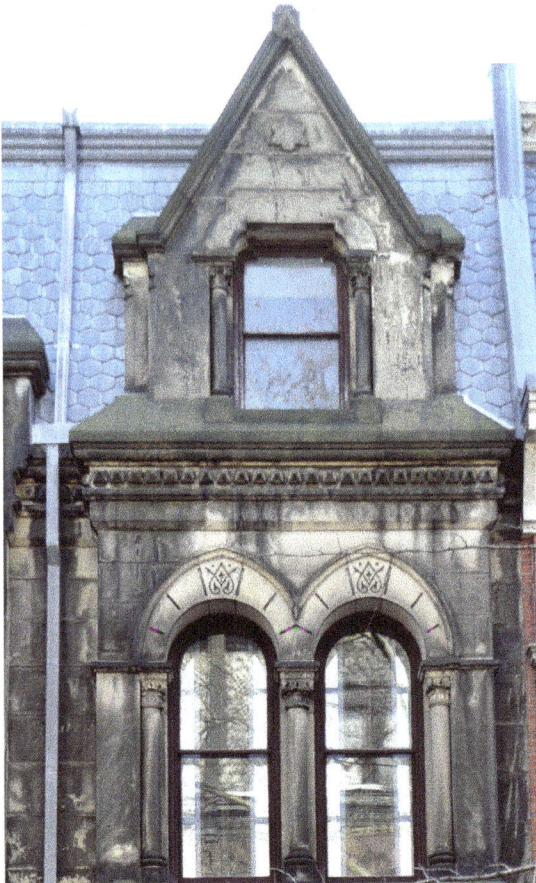

Biofilm. Cyanobacteria and fungi darken limestone façades.[12]

Biofilm from exhaust from a steam vent. It contains algae.

Bacteria have sex three ways: (1) genetic material (such as DNA) may pass from one bacterium to another through direct contact of the two organisms; (2) it may pass indirectly through a vector (such as a virus); or (3) it may pass as a free molecule through the environment.[2]

Rosemary Redfield, microbial geneticist and evolutionary biologist at the University of British Columbia, pointed out that bacterial ingestion of DNA is primarily nutritional. It is more like eating than like sex.[3] This interpretation points to a possible microbial pathway for evolution of sex. Bacteria have been credited with evolution of precursors to the cellular machinery of sex in more complex forms of life.[4]

Microorganisms stuck together on a surface form a biofilm. Their secretions hold the organisms in place.[5] Microbial communities in biofilms form microscopic ecosystems.[6] Experimentally, rates of bacterial mating were 1,000 times higher inside than outside a biofilm.[7] Bacterial sex pheromone promoted mating, increasing rates of gene (plasmid) transfer as much as 100,000 times.[8] Biofilms promoted action of bacterial sex pheromones.[9] Genes most commonly studied were those that confer resistance to antibiotics.

Bacterial sex may occur wherever biofilms occur, such as in the human gut.[10] In cities, bacterial sex has been documented in municipal sewage water treatment plants.[11] Bacterial sex may occur on biofilms coating façades of old buildings, but I have found no studies that explored this possibility.

Chapter 13 // Fungi, Bacteria, and Lichens

Vegetative Reproduction: lemon lichen (*Candelaria concolor*)

Sexual reproduction in a lichen must cope with a partnership of two or more organisms.

A lichen is a composite organism composed of one or more fungi plus one or more photosynthetic partners (algae or cyanobacteria).[1] Sexual reproduction in lichens conventionally refers only to the dominant fungus, which may produce sexual fungal structures, or fruiting bodies, that make sexually derived spores. Among lichens found in Great Britain and Ireland, those that produce sexual spores constitute 90 percent, whereas those that produce vegetative propagules constitute only 29 percent; many lichens produce both.[2]

When a lichen fungus reproduces sexually, it may have to secure not only a sexual partner but also a photosynthetic partner. Dispersal of a lichen sexual spore may separate the spore from the symbiotic community of microorganisms that constitute the lichen. Reconstitution of lichens requires habitats that support uniting a lichen fungus with its photosynthetic partner.[3] In cities, such habitats may be scarce. Some sexually reproducing lichens self-fertilize, avoiding the problem of finding a sexual partner.[4] Confronting many challenges, lichens have diversified their reproductive modes. One review of all these reproductive modes counted a total of twenty-seven.[5]

Lemon lichen is common in cities.[6] It illustrates how a lichen may successfully cope with urbanization. In downtown Philadelphia, I have found that its dominant reproductive structures are vegetative rather than sexual. Vegetative propagules (soredia) cover much of its surface. They enable the lichen to propagate without breaking up and reconstituting its fungal-algal partnership. Lemon lichen is reported to make uncommon minute sexual structures.[7] In the course of evolution, lichens have transitioned from asexual to sexual reproduction and vice versa.[8] Lemon lichen's current investment in vegetative reproduction may have contributed to its success propagating in cities.

Lemon lichen on street tree.

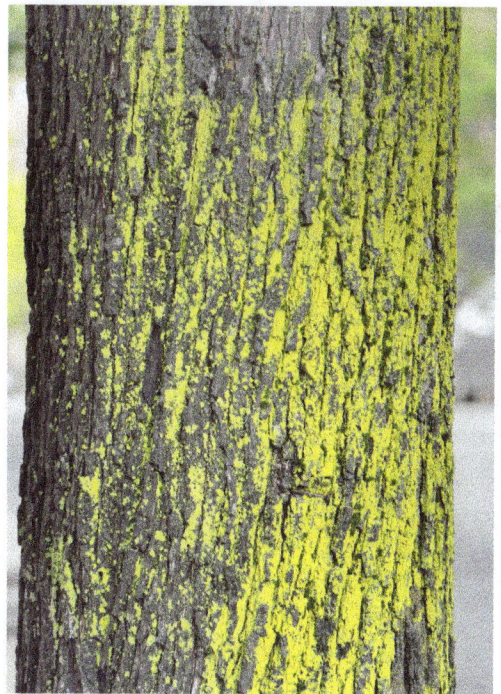

Close-up of tree shown above.

Mating Types: mushroom-forming fungi (Agaricomycetes)

Some mushroom-forming fungi have sex without sexes. They have mating types.

Sex in the inky cap mushroom-forming fungus *Coprinopsis cinerea* begins before the fruiting body, or mushroom, appears. The fungus first makes microscopic branching filaments, or hyphae, that form a mesh over its substrate. The hyphal network, or mycelium, supports both nutrition and mating.[1] When hyphae from two mycelia of this species meet, the intersecting pair of cells fuse, connecting the two hyphal networks; however, the two nuclei within the fused cell do not fuse. If the two mycelia happen to differ in mating type (i.e., are genetically compatible), each nucleus migrates into the opposite mycelium.[2] This initiates the process of mating. Fusion of nuclei descended from the two mycelia occurs later, in the mushroom. A distinctive feature of mating in mushroom-forming fungi is the delay in combining genomes of a mated pair.[3]

Gametes in a mating pair of mushroom-forming fungi are hyphae. They may be structurally identical, in contrast to eggs and sperm in plants and animals. In outcrossing species, each mycelium has a genetically determined mating type, and two mycelia can mate only if their mating types differ.[4] In contrast to sexes in most forms of life, mating types in a mushroom-forming fungus are not limited to two; they can number in the tens of thousands.[5] This large number of mating types boosts the chances that two random mycelia that happen to meet will constitute a genetically compatible sexual pair.[6] Diversity of mating types may facilitate mating of a mushroom-forming fungus in urban habitats where chances of meeting a potential mate are low.

Inky cap mushroom (unidentified coprinoid species) sprouting from wood chips beside a sidewalk downtown. These mushrooms appeared in the morning; by early afternoon, they had begun to wither. By the end of the day, all had toppled over and shriveled. Hyphae of mushroom-forming fungi initiate mating in substrate out of sight long before the mushroom appears. Completion of sexual union (fusion of nuclei) and production of spores happens later, in the mushroom. Ephemerality of this inky cap mushroom protects its final reproductive step from disturbance—a high risk in downtown Philadelphia.

Chapter 13 // Fungi, Bacteria, and Lichens

Microbial-Induced Change in Sex: pill bug (*Armadillidium vulgare*)

A bacterial infection causes pill bugs born genetically male to develop into functional females.

Pill bugs live in gardens, where they disappear under stones in the day and feed on decaying vegetable matter at night. They are crustaceans, which also include lobsters, shrimp, and crabs.

Infection can cause young, genetically male pill bugs to differentiate sexually into functional females. The infectious agent is a bacterium named *Wolbachia*. It is an endosymbiont, meaning it lives symbiotically inside cells of the pill bug. Infected female pill bugs transmit *Wolbachia* through their eggs to offspring. Male pill bugs can do the same after infection with *Wolbachia* has transformed them into functional females. By converting genetic males into functional females, *Wolbachia* increases its own infectivity. It behaves like a selfish gene. Populations of pill bugs infected by *Wolbachia* appear to be predominantly females.[1]

Pill bug (*Armadillidium vulgare*).

Most described species on our planet are arthropods.[2] In addition to crustaceans, arthropods include insects, spiders, and scorpions. *Wolbachia* infects an estimated half of all species of terrestrial arthropods.[3] It manipulates sexual biology in diverse ways depending on species of arthropod and strain of the bacterium.[4]

Pill bug (*A. vulgare*).

Infection of pill bugs by *Wolbachia* in our garden theoretically could protect pill bugs from reproductive failure. I periodically change the layout of our garden. This disturbance forces pill bugs to disperse to new territory and to found fresh populations. A solitary male pill bug that dispersed to a fresh part of the garden can neither bear young nor found a new population, but a fertilized male feminized by *Wolbachia* can accomplish both.

Pill bug (*A. vulgare*) curled up.

Sex for Biological Control: Asian tiger mosquito (*Aedes albopictus*)

Infection of male Asian tiger mosquitos with a particular strain of Wolbachia *in the laboratory blocks sexual reproduction of this pest in the field.*

The Asian tiger mosquito has invaded five continents, including much of North America. It has spread especially to cities and suburbs. In Philadelphia, it aggressively bites. Despite its potential as a vector of viruses causing human disease, it has largely defied conventional attempts at control.[1]

Bacteria classified as *Wolbachia* have been recruited for biological control of the Asian tiger mosquito. *Wolbachia* bacteria were first discovered in 1924, in the northern house mosquito, at the Harvard Medical School by Marshall Hertig and Simeon Wolbach.[2] In 2010, Maurizio Calvitti from the Italian National Agency for New Technologies and colleagues transferred a strain of this *Wolbachia* from the northern house mosquito to the Asian tiger mosquito. They then found that female Asian tiger mosquitoes without this strain of *Wolbachia* became sterile after mating with male Asian tiger mosquitoes with this strain.[3]

Female Asian tiger mosquito in our house.

Wolbachia-induced sterilization laid a foundation for biological control. Since 2019, reports have described successful field trials using *Wolbachia* infection for suppression of populations of the Asian tiger mosquito and its close relative, the yellow fever mosquito (*Aedes aegypti*). These trials have been conducted in China, California, and Florida. They have involved release of millions of *Wolbachia*-infected male mosquitoes (which do not bite).[4] Both these species of mosquito can transmit Zika and dengue viruses. In separate studies, *Wolbachia* infection of these two species has blocked transmission of both viruses.[5]

Current technology using *Wolbachia* to block reproduction in mosquitoes is evolving. Sustained suppression of mosquito abundance in particular locales will likely require repeated treatments. Whether *Wolbachia* will be widely deployed for biological control of mosquitoes and their viruses in cities remains to be seen.[6] Genetic engineering of *Aedes* mosquitoes offers an alternative technology for biological control.[7]

/ CONCLUSION /

Does Reproductive Diversity Matter?

Reproductive diversity is a source of beauty and wonder, and it is integral to biodiversity. What is its functional significance?

Fitness of some populations of common ragweed (page 17) and yellow toadflax (page 26) decreased with urbanization. Both species are self-incompatible cross-fertilizers.[1] In metropolitan Paris, a floral survey found shifts in the composition of communities of street plants with urbanization, toward species that are self-compatible and that self-fertilize.[2] In populations in diverse fragmented habitats around the globe, studies have suggested a shift to self-fertilization.[3] In Britain and the Netherlands, cross-fertilizing species of plants dependent on declining insect pollinators have themselves declined compared to other species of plants.[4]

How might we recognize species of street plants that predominantly self-fertilize? Self-fertilizing herbaceous plants other than grasses tend to have small floral displays.[5] Self-fertilizing annuals tend to have flowers that are both small and ephemeral, and the plants themselves tend to be short.[6] Examples illustrated in this book include common chickweed, prostrate knotweed, and carpetweed.[7]

Paris. With urbanization, composition of communities of street plants in metropolitan Paris shifted toward species that self-fertilize.

2004 in Paris, by Thomas Dahlstrøm Nielsen (CC BY-SA 4.0). Image cropped and exposure adjusted. URLs on page 162.

Short term, self-fertilization may rescue populations that cannot obtain mates. Long term, self-fertilization may deny these populations the evolutionary potential they require to adapt.[8] Will self-fertilizing populations in cities survive evolutionary arms races? Will they evolve fast enough to keep up with climate change? Observation of reproductive diversity in cities over time may help provide answers.

Glossary*

androgenesis A rare form of quasi-sexual reproduction in which sperm cells pass on all their genes to progeny and egg cells pass on almost none of theirs, as in the Asian clam (page 81).

anther In a flower, the male structure that makes pollen, as illustrated in chickweed (page 11) and jewelweed (page 13).

apomixis Production of seeds or spores without fertilization, as in North American common dandelion (page 23) and purple cliffbrake fern (page 6).[1]

asexual reproduction Propagation without fusion of gametes. Examples are vegetative reproduction, parthenogenesis, apomixis, and binary fission (in bacteria).

binary fission Reproduction by division of a body into two bodies, as in bacterial cells.

bisexual In plants, having structures and functions of both sexes, as illustrated in flowers of common chickweed (page 11); hermaphroditic.

brood parasite A mother who induces others to raise her young. Female brown-headed cowbirds (page 117) and lemon cuckoo bumblebees (page 52) are specialized brood parasites.

cross-fertilization Fertilization of one individual by another. All sexually reproducing insects and vertebrates presented in this book cross-fertilize.

cross-pollination Transfer of pollen from anthers (male parts) of flowers to stigmas (female parts) of flowers on different plants, usually of the same species, as in box elder (page 34).

cuckoldry Fertilization of the socially established mate of a male by another male. Satellite and sneaker bluegill sunfish are specialized cuckolders (page 93).

diploid A cell or organism having two sets of chromosomes, typically one set from each parent, as in female bumblebees.

epigenetic change Heritable change without a change in the sequence of genetic code (DNA), as in clones of common dandelion[2] (page 23).

evolutionary arms race A struggle between two adversarial species, each developing weapons and counterweapons against the other, as in white campion and its fungal parasite (page 20).

evolutionary potential Capacity for heritable change in a finite period of time. It is presumably low in clonal populations as in the Asian clam (page 81) and threadstalk speedwell (page 32).

extra-pair copulation Mating of an animal with an individual that is not its social mate. It is common in birds that are socially monogamous, such as the Canada goose (page 115).

femme fatale A female that preys on males that she sexually attracts, as in predatory fireflies (page 66) and cannibalistic female praying mantises (pages 63–64).

fertilization Fusion of nuclei of two gametes (such as an egg cell and sperm cell) to form a cell (zygote) with a nucleus having genetic material from both parental nuclei.

gamete A mature germ cell that unites with another mature germ cell during fertilization, as in egg cells and sperm cells. Hyphae of some fungi behave as gametes (page 122).

gametophyte The gamete-producing phase in the life cycle of mosses, ferns and liverworts, as in the case of a male umbrella liverwort (illustrated on page 3).

gene flow Mixing of genes of one population with another, as when a member of one population mates with a member of another in western black widow spiders (page 79).

genetic drift Chance swings in the prevalence of particular genes within small, isolated populations, as in urban ragweed[3] (page 17) and urban toads[4] (page 90).

genetic isolation Lack of gene flow of a population to or from other populations of the same species, as in the dusky salamander in Manhattan, New York City[5] (page xvi).

genetic recombination Reshuffling of genes through meiosis or other cellular processes, such as those noted in bacteria (page 120). It was undetectable in milkweed aphid clones[6] (page 72).

---

* This glossary includes some terms absent in the text but common in cited references.

gynogenesis A quasi-sexual form of reproduction in which egg cells pass on all their genes to future generations and sperm pass on none of theirs, as in silver cinquefoil[7] (page 24).

habitat fragmentation Division of habitat into pieces, such as urban gardens, lawns, vacant lots, ponds, and parks. It contributes to sexual isolation.

haploid An organism or cell having only one set of chromosomes, as in male bumblebees and gametophytes of mosses and liverworts.

haploid-diploid sex determination Sex determination in which fertilized (diploid) eggs make females, and unfertilized (haploid) eggs make males, exemplified in bees, ants, and wasps.[8]

hermaphrodite An individual that possesses structures and function of both sexes, as in a gray garden slug (page 82) and a jewelweed flower (page 13).

hybrid species A species created by hybridization of two species, as noted in urban salsifies (page 31), silver cinquefoil (page 24), and common chickweed (page 10).

hybrid vigor Superior structure or function of outcrossed offspring compared to their parents, as in hybrids of exotic and native subspecies of common reed (page 29).

inbreeding depression Reduced survival or fertility of offspring of closely related parents, as noted in mated German cockroach siblings (page 71) and in self-fertilized yellow toadflax.[9]

jumping genes Sequences of DNA that change position within a genome. Synonyms are transposons and transposable elements. They genetically modify dandelion clones[10] (page 23).

mating type The molecular genetic mechanism that controls self-incompatibility in flowering plants and fungi, as in yellow toadflax (page 26) and the fungus *Coprinopsis cinerea* (page 122).

matricide Killing the queen in a colony of social insects by her workers (daughters), as in the bald-faced hornet (page 50).

meiosis A specialized type of cell division used in production of gametes. It halves the number of chromosomes in cells and enables genetic recombination.

monandry The state of a female with only one mate, as in the fragile forktail damselfly (page 62), bald-faced hornet (page 50), and most queens of the common eastern bumblebee (page 51).

monogamy The state of having only one mate. Social monogamy in birds may last a lifetime or only a season. Canada geese (page 115) are socially monogamous but biologically polygamous.

outbreeding depression Reduced survival or fertility of offspring of parents who are genetically distantly related, as in hybrids of American and oriental bittersweet (page 37).

outcrossing Crossing of individuals that are not closely related. Self-incompatible plants predominantly outcross, as in yellow toadflax (page 26) and common ragweed (page 16).

panmixia Random mixing and mating within a population, as in the American eel. The global reproductive population of this eel mixes and spawns in the Sargasso Sea (page 91).

parthenogenesis Development of an embryo from an unfertilized egg. In this book the term refers exclusively to animals (such as aphids, page 72), but the term may apply also to plants.

pollination Transfer of pollen from the anther (male part) of a flower to the stigma (female receptor) of the same or a different flower, as in common chickweed (illustrated on page 11).

polyandry The state of a female having more than one mate. Female garter snakes (page 86), female house mice (page 100), and pavement ant queens may be polyandrous (page 56).

polygamy The state of having more than one mate. The term applies to either a male or a female, or both as in blue dashers (page 58), bedbugs (page 74), and house mice (page 100).

polygyny In the European starling (page 118), refers to the social bonding of a male to two or more females. Biologically polygamous males are not necessarily socially polygynous.

polyploid An organism or cell with more than two sets of chromosomes, as in North American dandelion (page 23), cinquefoil (page 24), and the earthworm *Octolasion tyrtaeum*[11] (page 83).

protandry Expression of male function before female function, as in chickweed (pages 10 and 11), the cabbage white butterfly (page 41), and the common garter snake (page 87).

protogyny Expression of female function before male function. In a Norway maple tree, female flowers bloom before male flowers, or vice versa (page 36).

self-compatible Having a mating system that allows self-fertilization, as in chickweed (page 10), wood sorrel (page 22), jewelweed (pages 13-14), and annual bluegrass (page xv).

self-fertilization Fusion of two gametes (such as an egg cell and a sperm cell) produced by the same individual, as in dwarf bristle-moss (page 2) and unmated gray garden slugs (page 82).

self-incompatible Having a molecular genetic mating system that blocks self-fertilization, as in threadstalk speedwell (page 32) and the fungus *Coprinopsis cinerea* (page 122).

self-pollination Transfer of pollen from the anther (male part) of a flower to the stigma (female part) of a flower on the same plant, as illustrated in chickweed (page 11).

sex-limited mimicry Mimicry that is limited to one sex but that is not sexual. An example is mimicry of the pipevine swallowtail by the black female eastern tiger swallowtail (pages 44-45).

sexual cannibalism Consumption of sexual rivals or suitors, as in praying mantises (pages 63-64), fireflies (page 66) and fragile forktail damselflies (page 62).

sexual isolation Mating greater within than between populations.[12] Sexual isolation is high between populations of northern house mosquitos above ground and below ground (page 70).

sexual mimicry Mimicry of one sex by the same or opposite sex, as in blue dashers (page 60), bluegills (page 93), fireflies (page 66), and praying mantises (page 64).

sexual reproduction Propagation that includes fusion of nuclei of gametes, such as those of egg cells and sperm cells or two hyphal cells.

sporophyte The spore-producing phase in the life cycle of mosses, liverworts, and ferns (illustrated on page 6). It contrasts with gametophyte.

stamen The male flower part. It consists of an anther and its supporting filament, illustrated in chickweed (page 11).

stigma In a flower, the female part that receives pollen, as illustrated in chickweed (page 11) and common ragweed (pages 15-16).

subsexual system A cellular system producing genetic recombination without fertilization, as in ferns[13] (page 6) and common dandelion[14] (page 23).

triploid An organism or cell having three sets of chromosomes, as in purple cliffbrake[15] (page 6), North American common dandelion[16] (page 23), and the Asian clam[17] (page 81).

vegetative reproduction Propagation without fertilization and typically involving multicellular structures such as (in plants) propagules, rhizomes, stolons, roots, rootlets, and stems.

zygote A fertilized cell formed by fusion of nuclei of two gametes, such as an egg cell and sperm cell. Its nucleus contains genetic material from nuclei of both gametes.

Notes

Introduction

1. Aronson MFJ et al. 2014. A global analysis of the impacts of urbanization on bird and plant diversity reveals key anthropogenic drivers. *Proceedings of the Royal Society B* 281:20133330. https://doi.org/10.1098/rspb.2013.3330.

2. Hitchings SP, Beebee TJC. 1998. Loss of genetic diversity and fitness in common toad (*Bufo bufo*) populations isolated by inimical habitat. *Journal of Evolutionary Biology* 11:269–283; Dornier A, Cheptou P. 2012. Determinants of extinction in fragmented plant populations: *Crepis sancta* (Asteraceae) in urban environments. *Oecologia* 169:703–712.

3. Hutchinson CS, Seymour GB. 1982. *Poa Annua* L. *Journal of Ecology* 70:887–901.

4. Wright SI, Kalisz S, Slotte T. 2013. Evolutionary consequences of self-fertilization in plants. *Proceedings of the Royal Society B* 280:20130133. https://doi.org/10.1098/rspb.2013.0133.

5. Clifford HT. 1956. Seed dispersal on footwear. *Proceedings of the Botanical Society of the British Isles* 2:129–131.

6. Schmidt W. 1989. Plant dispersal by motor cars. *Vegetatio* 80:147–152.

7. Dindal DL, ed. 1990. *Soil Biology Guide.* New York: John Wiley & Sons.

8. Bergey EA, Figueroa LL, Mather CM, Martin RJ, Ray EJ, Kurien JT, Westrop DR, Suriyawong P. 2014. Trading in snails: plant nurseries as transport hubs for non-native species. *Biological Invasions* 16:1441–1451.

9. Carlen E, Munshi-South J. 2020. Widespread genetic connectivity of feral pigeons across the northeastern megacity. *Evolutionary Applications* 2020:1–13. https://doi.org/10.1111/eva.12972; Henger CS, Herrera GA, Nagy CM, Weckel ME, Gormezano LJ, Wultsch C, Munshi-South J. 2020. Genetic diversity and relatedness of a recently established population of eastern coyotes (*Canis latrans*) in New York City. *Urban Ecosystems* 23:319–330; Combs M, Puckett EE, Richardson J, Mims D, Munshi-South J. 2018. Spatial population genomics of the brown rat (*Rattus norvegicus*) in New York City. *Molecular Biology* 27:83–98; Munshi-South J. 2012. Urban landscape genetics: canopy cover predicts gene flow between white-footed mouse (*Peromyscus leucopus*) populations in New York City. *Molecular Ecology* 21:1360–1378.

10. Munshi-South J, Zak Y, Pehek E. 2013. Conservation genetics of extremely isolated urban populations of the northern dusky salamander (*Desmognathus fuscus*) in New York City. *Peer J* 1:e64. https://doi.org/10.7717/peerj.64.

11. Brown A. 1880. Ballast plants in and near New York City. *Bulletin of the Torrey Botanical Club* 7:122–126.

12. Rutz GM, Carlton JT, Grosholz ED, Hines AH. 1997. Global invasions of marine and estuarine habitats by non-indigenous species: mechanisms, extent and consequences. *American Zoologist* 37:621–632.

13. Mills EL, Leach JH, Carlton JT, Secor CL. 1994. Exotic species and the integrity of the Great Lakes: Lessons from the past. *Bioscience* 44:666–676.

14. Perillo JA, Butler LH. 2009. Evaluating the use of Fairmount Dam fish passage facility with application to anadromous fish restoration in the Schuylkill River, Pennsylvania. *Journal of the Pennsylvania Academy of Science* 83:24–33.

15. Villaseñora NR, Chianga LA, Hernándeza HJ, Escobar MAH. 2020. Vacant lands as refuges for native birds: An opportunity for biodiversity conservation in cities. *Urban Forestry and Urban Greening* 49:126632. https://doi.org/10.1016/j.ufug.2020.126632; Turo KJ, Spring MR, Sivakoff FS, Delgado de la flor YA, Gardiner MM. 2020. Conservation in post-industrial cities: How does vacant land management and landscape configuration influence urban bees? *Journal of Applied Ecology* 58:58–69.

16. Löki V, Deák B, Lukács AB, V. AM. 2019. Biodiversity potential of burial places—a review on the flora and fauna of cemeteries and churchyards. *Global Ecology and Conservation* 18:e00614. https://doi.org/10.1016/j.gecco.2019.e00614.

17. Kowarik I, von der Lippe M. 2011. Secondary wind dispersal enhances long-distance dispersal of an invasive species in urban road corridors. *NeoBiota* 9:49–70; Miller NG, McDaniel SF. 2004. Bryophyte dispersal inferred from colonization of an introduced substratum on Whiteface Mountain, New York. *American Journal of Botany* 91 1173–1182.

18. Trombulak SC, Frissell CA. 2000. Review of ecological effects of roads on terrestrial and aquatic communities. *Conservation Biology* 14:18–30.

19. Mühlenbach V. 1979. Contributions to the synanthropic (adventive) flora of the railroads in St. Louis, Missouri, U.S.A. *Annals of the Missouri Botanical Garden* 66:1–108.

20. Owens ACS, Lewis SM. 2018. The impact of artificial light at night on nocturnal insects: A review and synthesis. *Ecology and Evolution* 2018:11337-11358; Sanders D, Frago E, Kehoe R, Patterson C, Gaston KJ. 2021. A meta-analysis of biological impacts of artificial light at night. *Nature, Ecology & Evolution* 5:74–81.

21. Belliard J, Beslagic S, Tales E. 2021. Changes in fish communities of the Seine basin over a long-term perspective. In *The Seine River Basin*, edited by Flipo N, Labadie P, Lestel L, 301-322. Chan, Switzerland: Springer.

Page 2. Spores: dwarf bristle-moss (*Orthotrichum pumilum*)

1. Fudali E. 2018 Expansion of epiphytic moss *Orthotrichum pumilum* (Orthotrichaceae) in Wrocław. *Fragmenta Floristica et Geobotanica Polonica* 25:295–298.

2. Crum HA, Anderson LE. 1981. *Mosses of Eastern North America* in 2 volumes. New York: Columbia University Press.

3. Crum and Anderson, *Mosses of Eastern North America*; Frahm J-P. 2007. Diversity, dispersal and biogeography of bryophytes (mosses). *Journal Biodiversity and Conservation* 17:277–284.

4. Miles CJ, Longton RE. 1992. Deposition of moss spores in relation to distance from parent gametophytes. *Journal of Bryology* 17:355–368.

5. Vanderpoorten A, Patiño J, Désamoré A, Laenen B, Górski P, Papp B, Holá E, Korpelainen H, Hardy O. 2019. To what extent are bryophytes efficient dispersers? *Journal of Ecology* 107:2149-2154; Patiño J, Vanderpoorten A. 2018. Bryophyte biogeography. *Critical Reviews in Plant Sciences*. https://doi.org/10.1080/07352689.2018.1482444; Miller NG, McDaniel SF. 2004. Bryophyte dispersal inferred from colonization of an introduced substratum on Whiteface Mountain, New York. *American Journal of Botany* 91:1173–1182.

Page 3. Cross-Fertilization: umbrella liverwort (*Marchantia polymorpha*)

1. Schuster RM. 1992. *Marchantia polymorpha* L. In *The Hepaticae and Anthocerotae of North America East of the Hundredth Meridian*, vol. VI, 324–334. Chicago: Field Museum of Natural History.

2. Pressel S, Duckett JG. 2019. Do motile spermatozoids limit the effectiveness of sexual reproduction in bryophytes? Not in the liverwort *Marchantia polymorpha. Journal of Systematics and Evolution* 57 371–381.

Page 4. Vegetative Reproduction: umbrella liverwort (*Marchantia polymorpha*)

1. Brodie HJ. 1951. The splash-cup dispersal mechanism in plants. *Canadian Journal of Botany* 29:224–234.

2. Tarén N. 1958. Factors regulating the initial development of gemmae in *Marchantia polymorpha. The Bryologist* 61:191–204.

3. Une K. 1984. A field observation on the reproductive mode in *Marchantia polymorphia* L. *Hikobia* 9:15–18.

4. Joenje W, During HJ. 1977. Colonisation of a desalinating Wadden-polder by bryophytes. *Vegetatio* 35:177–185.

5. Pressel S, Duckett JG. 2019. Do motile spermatozoids limit the effectiveness of sexual reproduction in bryophytes? Not in the liverwort *Marchantia polymorpha. Journal of Systematics and Evolution* 57 371–381.

6. McLetchie DN, Puterbaugh MN. 2000. Population sex ratios, sex specific clonal traits and tradeoffs among these traits in the liverwort *Marchantia inflexa. Oikos* 90:227–237.

7. Schofield WB, Crum HA. 1972. Disjunctions in bryophytes. *Annals of the Missouri Botanical Garden* 59:174–202.

Page 5. Self-Fertilization: hemispheric liverwort (*Reboulia hemisphaerica*)

1. Schuster RM. 1992. *Reboulia hemisphaerica* (L) Raddi. In *The Hepaticae and Anthocerotae of North America East of the Hundredth Meridian*, vol. VI, 152–163. Chicago: Field Museum of Natural History.

2. Dupler AW. 1922. The male receptacle and antheridium of *Reboulia hemisphaerica. American Journal of Botany* 9: 285–295.

3. Haig D. 2016. Living together and living apart: the sexual lives of bryophytes. *Philosophical Transactions of the Royal Society B* 371:20150535. https://doi.org/10.1098/rstb.2015.0535.

Page 6. Spores Without Fertilization: purple cliffbrake fern (*Pellaea atropurpurea*)

1. Nayar BK, Kaur S. 1971. Gametophytes of homosporous ferns. *Botanical Review* 37:295–396.

2. Steil WN. 1911. Apogamy in *Pellaea atropurpurea*. *Botanical Gazette* 52:400–401.

3. Hayes DW. 1924. Some studies of apogamy in *Pellaea atropurpurea* (L.) Link. *Transactions of the American Microscopical Society* 43:119–135.

4. Ishikawa H, Ito M, Watano Y, Kurita S. 2003. Electrophoretic evidence for homoeologous chromosome pairing in the apogamous fern species *Dryopteris nipponensis* (Dryopteridaceae). *Journal of Plant Research* 116:165–167.

5. Klekowski EJ Jr. 1973. Sexual and subsexual systems in homosporous pteridophytes: a new hypothesis. *American Journal of Botany* 60:535–544; Tryon AF. 1972. Spores, chromosomes and relations of the fern *Pellaea atropurpurea*. *Rhodora* 74:220–241.

6. Wagner WJ, Johnson D. 1981. Natural history of the ebony spleenwort, *Asplenium platyneuron* (Aspleniaceae), in the Great Lakes area. *Canadian Field-Naturalist* 95:156–166; Crist KC, Farrar DR. 1983. Genetic load and long-distance dispersal in *Asplenium platyneuron*. *Canadian Journal of Botany* 61:1809–1814.

Page 7. Lost Sex: marble screw-moss (*Syntrichia papillosa*)

1. Giudice RL. 1994. A new record for the Sicilian bryoflora: *Tortula papillosa* (Pottiaceae). *Flora Mediterranea* 4:37–39.

2. Trigoboff N. 2005. *Tortula papillosa* and *Tortula pagorum* (Pottiaceae) in New York State. *Evanasia* 22:85–89; Floyed A, Gibson M. 2012. Bryophytes of urban industrial streetscapes in Victoria, Australia. *Victorian Naturalist* 129:203–214.

3. Crum HA, Anderson LE. 1981. *Mosses of Eastern North America* in 2 volumes. New York: Columbia University Press; Catcheside DG. 1980. *Mosses of South Australia*. Adelaide, South Australia: The Flora and Fauna of South Australia Handbooks Committee; Mishler BD. 2007. *Syntrichia papillosa*. In *Flora of North America*. Volume 27, edited by Flora of North America Editorial Committee, 619-621. New York and Oxford: Oxford University Press.

4. Duckett J, Ligrone R. 1992. A survey of diaspore liberation mechanisms and germination patterns in mosses. *Journal of Bryology* 17:335–354.

5. Mischler BD. 1988. Reproductive ecology of bryophytes. In *Plant Reproductive Ecology: Patterns and Strategies*, edited by Doust JL, Doust LL, 285–306. Oxford: Oxford University Press.

Page 8. Sporadic Sex: silver bryum moss (*Bryum argenteum*)

1. Cooke WB. 1953. Mosses in a sewage treatment plant. *The Bryologist* 56:143-145; Porley R, Hodgetts N. 2005. *Mosses and Liverworts*. London: Collins; Rusinska A, Balcerkiewicz S. 1979. Moss communities on the roofs of buildings. *Abstracta Botanica* 5:51–60.

2. Clare D, Terry T. 1960. Dispersal of *Bryum argenteum*. *Transactions of the British Bryological Society* 3:748;

3. Selkirk PM, Skotnicki ML, Ninham J, Connett MB, Armstrong J. 1998. Genetic variation and dispersal of *Bryum argenteum* and *Hennediella heimii* populations in the Garwood Valley, southern Victoria Land, Antarctica. *Antarctic Science* 10:423–430.

4. Raudenbush Z, Greenwood J, McLetchie D, Eppley S, Keeley S, Castetter R, Stark L. 2018. Divergence in life-history and developmental traits in silvery-thread moss (*Bryum argenteum* Hedw.) genotypes between golf course putting greens and native habitats. *Weed Science* 66:642–650.

5. Pôrto KC, Silva ICCe, Reis LCd, Maciel-Silva AS. 2017. Sex ratios and sporophyte production in the moss *Bryum argenteum* Hedw. on a rock outcrop, north-eastern Brazil. *Journal of Bryology* 39:194–198.

Page 9. Chapter 2 Herbaceous Annual Plants

1. Snell R, Aarssen LW. 2005. Life history traits in selfing versus outcrossing annuals: Exploring the 'time-limitation' hypothesis for the fitness benefit of self-pollination. *BMC Ecology* 5. https://doi.org/10.1186/1472-6785-5-2.

Page 10. Opportunistic Cross-Fertilization: common chickweed (*Stellaria media*)

1. Sobey DG. 1981. *Stellaria media* (L.) Vill. *Journal of Ecology* 69:311–335.

2. Müller H. 1883. *The Fertilization of Flowers*, translated, compiled, and edited by D'Arcy W. Thompson with a preface by Charles Darwin. London: Macmillan and Company.

3. Salisbury EJ. 1974. The variations in the reproductive organs of *Stellaria media* (sensu stricto) and allied species with special regard to their relative frequency and prevalent modes of pollination. *Proceedings of the Royal Society of London. Series B* 185:331–342; Whitehead F, Sinha R. 1967. Taxonomy and taxometrics of *Stellaria media* (L.) Vill., *S. neglecta* Weihe and *S. pallida* (Dumort.). *New Phytologist* 66:769–784; Turkington R, Kenkel NC, Franko GD. 1980. The biology of Canadian weeds: 42. *Stellaria media* (L.) Vill. *Canadian Journal of Plant Science* 60:981–992.

4. Schmidt W. 1989. Plant dispersal by motor cars. *Vegetatio* 80:147–152; Clifford HT. 1956. Seed dispersal on footwear. *Proceedings of the Botanical Society of the British Isles* 2:129–131; Salisbury E. 1964. *Weeds & Aliens*. London: Collins.

5. Darwin C. 1876. *The Effects of Cross and Self Fertilisation in the Vegetable Kingdom*. London: John Murray.

6. Morton JK. 2005. Documented chromosome numbers in North American species of *Silene* and *Stellaria* Caryophyllaceae). *Rhodora* 21:1669–1674; Sinha RP, Whitehead FH. 1965. Meiotic studies of British populations of *Stellaria media* (L.) Vill., *S. neglecta* Weihe and *S. pallida* (Dumort.) Piré. *New Phytologist* 64:343–345; Darmency H, Gasquez J. 1997. Spontaneous hybridization of the putative ancestors of the allotetraploid *Poa annua*. *New Phytologist* 136:497–501.

7. Soltis PS, Soltis DE. 2000. The role of genetic and genomic attributes in the success of polyploids. *Proceedings of the National Academy of Sciences* 97:7051–7057.

8. Byers DL, Waller DM. 1999. Do plant populations purge their genetic load? Effects of population size and mating history on inbreeding depression. *Annual Review of Ecology and Systematics* 30:479–513.

9. Strykstra RJ. 1989. Thrips as visitors and pollinators of *Stellaria media* L. (Abstract of paper presented at a meeting of the Netherlands Society for Plant Cell and Tissue Culture on 11 March 1988—A Meeting of the Royal Botanical Society of the Netherlands). *Acta Botanica Neerlandica* 38:101.

Page 12. Nectar Robbery: jewelweed (*Impatiens capensis*)

1. Rust RW. 1979. Pollination of *Impatiens capensis*: Pollinators and nectar robbers. *Journal of the Kansas Entomological Society* 52:297–308.

2. Zimmerman M, Cook S. 1985. Pollinator foraging, experimental nectar-robbing and plant fitness in *Impatiens capensis*. *American Midland Naturalist* 113:84–91.

3. Young HJ. 2008. Selection on spur shape in *Impatiens capensis*. *Oecologia* 156:535–543.

Page 13. Male-to-Female Sex Change: jewelweed (*Impatiens capensis*)

1. Rust RW. 1977. Pollination in *Impatiens capensis* and *Impatiens pallida* (Balsaminaceae). *Bulletin of the Torrey Botanical Club* 104:361–367.

2. Mitchell-Olds T, Waller DM. 1985. Relative performance of selfed and outcrossed progeny in *Impatiens capensis*. *Evolution* 1985:533–544.

3. Temeles EJ, Pan IL. 2002. Effect of nectar robbery on phase duration, nectar volume, and pollination in a protandrous plant. *International Journal of Plant Sciences* 163:803–808.

Page 14. Flowers That Never Open: jewelweed (*Impatiens capensis*)

1. Schemske DW. 1978. Evolution of reproductive characteristics in *Impatiens* (Balsaminaceae): The significance of cleistogamy and chasmogamy. *Ecology* 59:596–613.

2. Mitchell-Olds T, Bergelson J. 1990. Statistical genetics of an annual plant, *Impatiens capensis*. I. Genetic basis of quantitative variation. *Genetics* 124:407–415.

3. Waller DM. 1984. Differences in fitness between seedlings derived from cleistogamous and chasmogamous flowers in *Impatiens capensis*. *Evolution* 38(2):427–440; Mitchell-Olds T, Waller DM. 1985. Relative performance of selfed and outcrossed progeny in *Impatiens capensis*. *Evolution* 1985:533–544.

4. Waller DM. 1980. Environmental determinants of outcrossing in *Impatiens capensis* (Balsaminaceae). *Evolution* 34:747–761.

5. Trent JA. 1942. Studies pertaining to the life history of *Specularia perfoliata* (L.).A.DC., with special reference to cleistogamy. *Transactions of the Kansas Academy of Science* 45:152–164; Lord EM. 1980. Intra-inflorescence variability in pollen/ovule ratios in the cleistogamous species *Lamium amplexicaule* (Labiatae). *American Journal of Botany* 67:529–533; Solbrig OT, Newell SJ, Kincaid DT. 1980. The population biology of the genus *Viola*: I. The demography of *Viola sororia*. *Journal of Ecology* 68:521–546; Keeler KH. 1978. Intra-population differentiation in annual plants. II. Electrophoretic variation in

Veronica peregrina. *Evolution* 32:638–645; Costea M, Tardif FJ. 2005. The biology of Canadian weeds. 131. *Polygonum aviculare* L. *Canadian Journal of Plant Science* 85:481–506.

Page 15. Quantitative Gender: common ragweed (*Ambrosia artemisiifolia*)

1. Lloyd DG. 1980. Sexual strategies in plants. III. A quantitative method for describing the gender of plants. *New Zealand Journal of Botany* 18:103–108.

2. McKone MJ, Tonkyn DW. 1986. Intrapopulation gender variation in common ragweed (Asteraceae: *Ambrosia artemisiifolia* L.), a monoecious, annual herb. *Oecologia* 70:63–67.

3. Paquin V, Aarssen L. 2004. Allometric gender allocation in *Ambrosia artemisiifolia* (Asteraceae) has adaptive plasticity. *American Journal of Botany* 91:430–438.

4. Tanurdzic M, Banks JA. 2004. Sex-determining mechanisms in land plants. *The Plant Cell* vol. 16, S61–S71.

Page 16. Male-Male Competition: common ragweed (*Ambrosia artemisiifolia*)

1. Friedman J, Barrett SCH. 2011. The evolution of ovule number and flower size in wind-pollinated plants. *The American Naturalist* 177:246–257.

2. Martin MD, Chamecki M, Brush GS. 2010. Anthesis synchronization and floral morphology determine diurnal patterns of ragweed pollen dispersal. *Agricultural and Forest Meteorology* 150:1307–1317.

3. Friedman J, Barrett SCH. 2009. Wind of change: New insights on the ecology and evolution of pollination and mating in wind-pollinated plants. *Annals of Botany* 103:1515–1527.

4. Durham OC. 1935. The pollen content of the air in North America. *Journal of Allergy* 6:128–149.

5. Cruden R. 2000. Pollen grains: Why so many? *Plant Systematics and Evolution* 222:143–165.

Page 17. Urban Sexual Isolation: common ragweed (*Ambrosia artemisiifolia*).

1. Gorton AJ, Moeller DA, Tiffin P. 2018. Little plant, big city: A test of adaptation to urban environments in common ragweed (*Ambrosia artemisiifolia*). *Proceedings of the Royal Society B* 285:20180968. https://doi.org/10.1098/rspb.2018.0968.

2. Willi Y, Van Buskirk J, Hoffmann AA. 2006. Limits to the adaptive potential of small populations. *Annual Review of Ecology, Evolution, and Systematics* 37:433–458.

3. Martin M, Zimmer E, Olsen M. 2014. Herbarium specimens reveal a historical shift in phylogeographic structure of common ragweed during native range disturbance. *Molecular Ecology* 23:1701–1716.

4. City of Minneapolis. 2011. An Ordinance of the City of Minneapolis Amending Title 11, Chapter 227 of the Minneapolis Code of Ordinances Relating to Health and Sanitation: Nuisances Generally. Sect. 227.90 (amended). http://www.minneapolismn.gov/www/groups/public/@council/documents/webcontent/convert_261255.pdf.

Page 18. Evolutionary Arms Race: field dodder (*Cuscuta campestris*)

1. Dinelli G, Bonetti A, Tibiletti E. 1993. Photosynthetic and accessory pigments in *Cuscuta campestris* Yuncker and some host species. *Weed Research* 33:253–260.

2. Costea M, Tardif FJ. 2006. The biology of Canadian weeds. 133. *Cuscuta campestris* Yuncker, *C. gronovii* Willd. ex Schult., *C. umbrosa* Beyr. ex Hook., *C. epithymum* (L.) L. and *C. epilinum* Weihe. *Canadian Journal of Plant Science* 86:293–316.

3. Runyon JB, Mescher MC, Moraes CMD. 2006. Volatile chemical cues guide host location and host selection by parasitic plants. *Science* 313:1964–1967; Benvenuti S, Dinelli G, Bonetti A, Catizone P. 2005. Germination ecology, emergence and host detection in *Cuscuta campestris*. *Weed Research* 45:270–278.

4. Koch AM, Binder C, Sanders IR. 2004. Does the generalist parasitic plant *Cuscuta campestris* selectively forage in heterogeneous plant communities? *New Phytologist* 162:147–155.

5. Kim G, Westwood JH. 2015. Macromolecule exchange in *Cuscuta*–host plant interactions. *Current Opinion in Plant Biology* 26:20–25.

6. Shahid S et al. 2018. MicroRNAs from the parasitic plant *Cuscuta campestris* target host messenger RNAs. *Nature* 553:82–85.

Page 19. Chapter 3 Herbaceous Perennial Plants

1. Sparrow FK, Pearson NL. 1948. Pollen compatibility in *Asclepias syriaca*. *Journal of Agricultural Research* 77:187–199.

Page 20. Sexually Transmitted Disease: white campion (*Silene latifolia*)

1. Uva RH, Neal JC, DiTomaso JM. 1997. *Weeds of the Northeast.* Ithaca, NY: Cornell University Press.

2. Renner SS. 2014. The relative and absolute frequencies of angiosperm sexual systems: Dioecy, monoecy, gynodioecy, and an updated online database. *American Journal of Botany* 101:1588–1596.

3. Alexander HM, Antonovics J. 1988. Disease spread and population dynamics of anther-smut infection of *Silene alba* caused by the fungus *Ustilago violacea. Journal of Ecology* 76:91–104.

4. Gladieux P, Vercken E, Fontaine M, Hood M, Jonot O, Couloux A, Giraud T. 2011. Maintenance of fungal pathogen species that are specialized to different hosts: Allopatric divergence and introgression through secondary contact. *Molecular Biology and Evolution* 28:459–471.

5. Altizer SM, Thrall PH, Antonovics J. 1998. Vector behavior and the transmission of anther-smut infection in *Silene alba. The American Midland Naturalist* 139:147–163.

6. Schäfer AM, Kemler M, Bauer R, Begerow D. 2010. The illustrated life cycle of *Microbotryum* on the host plant *Silene latifolia. Botany-Botanique* 88:875–885.

7. Uchida W, Matsunaga S, Sugiyama R, Kazama Y, Kawano S. 2003. Morphological development of anthers induced by the dimorphic smut fungus *Microbotryum violaceum* in female flowers of the dioecious plant *Silene latifolia. Planta* 218:240–248.

8. Kawamoto H, Yamanaka K, Koizumi A, Hirata A, Kawano S. 2017. Cell death and cell cycle arrest of *Silene latifolia* stamens and pistils after *Microbotryum lychnidis-dioicae* infection. *Plant and Cell Physiology* 58:320–332.

9. Toh S, Chen Z, Rouchka E, Schultz D, Cuomo C, Perlin M. 2018. *Pas de deux*: An intricate dance of anther smut and its host. *G3 Genes/Genomes/Genetics* 8:505–518.

10. Fontaine MC, Gladieux P, Hood ME, Giraud T. 2013. History of the invasion of the anther smut pathogen on *Silene latifolia* in North America. *The New Phytologist* 198:946–956.

11. Giraud T, Jonot O, Shykoff J. 2005. Selfing propensity under choice conditions in a parasitic fungus, *Microbotryum violaceum*, and parameters influencing infection success in artificial inoculations. *International Journal of Plant Sciences* 166:649–657.

12. Clay K, Kover PX. 1996. The red queen hypothesis and plant/pathogen interactions. *Annual Review of Phytopathology* 34:29–50.

Page 21. Nocturnal Fertilization: white campion (*Silene latifolia*)

1. Baker HG. 1948. The ecotypes of *Melandrium dioicum* (L. Emend.) Coss. & Germ. *New Phytologist* 47:131–145; Medler JT. 1962. Morphometric studies on bumble bees. *Annals of the Entomological Society of America* 55:212–218.

2. Altizer SM, Thrall PH, Antonovics J. 1998. Vector behavior and the transmission of anther-smut infection in *Silene alba. The American Midland Naturalist* 139:147–163; Hawkins RP. 1969. Length of tongue in a honey bee in relation to the pollination of red clover. *Journal of Agricultural Science* 73:489–493.

3. Dötterl S, Jürgens A, Seifert K, Laube T, Weissbecker B, Schütz S. 2006. Nursery pollination by a moth in *Silene latifolia*: The role of odours in eliciting antennal and behavioural responses. *New Phytologist* 169:707–718; Waelti MO, Muhlemann JK, Widmer A, Schiestl FP. 2008. Floral odour and reproductive isolation in two species of *Silene. Journal of Evolutionary Biology* 21:111–121.

4. Young HJ. 2002. Diurnal and nocturnal pollination of *Silene alba* (Caryophyllaceae). *American Journal of Botany* 89:433–440.

5. Aonuma W, Shimizu Y, Ishii K, Fujita N, Kawano S. 2013. Maturation timing of stamens and pistils in the dioecious plant *Silene latifolia. Journal of Plant Research* 126:105–112.

6. Barthelmess EL, Richards CM, McCauley DE. 2006. Relative effects of nocturnal vs diurnal pollinators and distance on gene flow in small *Silene alba* populations. *New Phytologist* 169:689–698.

7. Jürgens A, Witt T, Gottsberger G. 1996. Reproduction and pollination in central European populations of *Silene* and *Saponaria* species. *Botanica Acta* 109:316–324.

8. Knop E, Zoller L, Ryser R, Gerpe C, Hörler M, Fontaine C. 2017. Artificial light at night as a new threat to pollination. *Nature* 548:206–209; Macgregor CJ, Evans DM, Fox R, Pocock MJO. 2017. The dark side of street lighting: Impacts on moths and evidence for the disruption of nocturnal pollen transport. *Global Change Biology* 23:697–707.

9. Altizer et al. Vector behavior and the transmission of anther-smut infection in *Silene alba.*

Page 22. Urban Self-Fertilization: creeping wood sorrel (*Oxalis corniculata*)

1. Shibaike H, Ishiguri Y, Kawano S. 1995. Reproductive biology of *Oxalis corniculata* (Oxalidaceae):

Style length polymorphisms and breeding systems of Japanese populations. *Plant Species Biology* 10:83–93; Shibaike H, Ishiguri Y, Kawano S. 1996. Population differentiation in floral and life history traits of *Oxalis corniculata* L. (Oxalidaceae) with style length polymorphism. *Journal of Plant Research* 109:315–325.

2. Doust LL, Doust JL, Cavers PB. 1981. Fertility relationships in closely related taxa of *Oxalis*, section Corniculatae. *Canadian Journal of Botany* 59:2603–2609.

3. Fukatsu M, Horie S, Maki M, Dohzono I. 2019. Hybridization, coexistence, and possible reproductive interference between native *Oxalis corniculata* and alien *O. dillenii* in Japan. *Plant Systematics and Evolution* 305:127–137; Soltis PS, Soltis DE. 2000. The role of genetic and genomic attributes in the success of polyploids. *Proceedings of the National Academy of Sciences* 97:7051–7057.

4. Groom QJ, Van der Straeten J, Hoste I. 2019. The origin of *Oxalis corniculata* L. *Peer J* 7:e6384. https://doi.org/10.7717/peerj.6384.

Page 23 Seeds Without Fertilization: common dandelion (*Taraxacum officinale*)

1. Iaffaldano B, Zhang Y, Cardina J, Cornis K. 2017. Genome size variation among common dandelion accessions informs their mode of reproduction and suggests the absence of sexual diploids in North America. *Plant Systematics and Evolution* 303:719–725.

2. Solbrig T. 1971. The population biology of dandelions. *American Scientist* 59:686–694.

3. King LM. 1993. Origins of genotypic variation in North American dandelions inferred from ribosomal DNA and chloroplast DNA restriction enzyme analysis. *Evolution* 47:136–151.

4. Meirmans P. 2005. *Ecological and Genetic Interactions Between Diploid Sexual and Triploid Apomictic Dandelions.* Dissertation. University of Amsterdam.

5. Tas ICQ, Van Dijk P. 1999. Crosses between sexual and apomictic dandelions (*Taraxacum*). I.: The inheritance of apomixis. *Heredity* 83:707–714.

6. Stebbins GL Jr. 1941. Apomixis in the angiosperms. *Botanical Review* 7:507–542.

7. Wilschut RA, Oplaat C, Snoek LB, Kirschner J, Verhoeven KJF. 2016. Natural epigenetic variation contributes to heritable flowering divergence in a widespread asexual dandelion lineage. *Molecular Ecology* 25:1759–1768; de Carvalho JF, Oplaat C, Pappas N, Derks M, de Ridder D, Verhoeven KJF. 2016. Heritable gene expression differences between apomictic clone members in *Taraxacum officinale*: Insights into early stages of evolutionary divergence in asexual plants. *BMC Genomics* 17:203; van Baarlen P, van Dijk P, Hoekstra R, de Jong J. 2000. Meiotic recombination in sexual diploid and apomictic triploid dandelions (*Taraxacum officinale* L.). *Genome* 43:827-835.

8. Van Dijk P, de Jong H, Vijverberg K, Biere A. 2009. An apomixis-gene's view on dandelions. In *Lost Sex. The Evolutionary Biology of Parthenogenesis*, edited by Schön I, Martens K, Van Dijk P, 475–493. Dordrecht, the Netherlands: Springer Netherlands.

Page 24. Gynogenesis: cinquefoil (*Potentilla argentea*)

1. Russell SD. 1993. The egg cell: Development and role in fertilization and early embryogenesis. *Plant Cell* 5:1349–1359.

2. Müntzing A. 1928. Pseudogamie in der gattung *Potentilla*. *Hereditas* 2:267–283.

3. Holm S, Ghatnekar L. 1996. Sexuality and no apomixis found in crossing experiments with diploid *Potentilla argentea*. *Hereditas* 125:77–82.

4. Mulligan GA. 1958. Chromosome numbers of Canadian weeds. II. *Canadian Journal of Botany* 37:81–92.

5. Holm S, Ghatnekar L. 1996. Apomixis and sexuality in hexaploid *Potentilla argentea*. *Hereditas* 125:53–60.

6. Paule J, Sharbel TF, Dobeš C. 2011. Apomictic and sexual lineages of the *Potentilla argentea* L. group (Rosaceae): Cytotype and molecular genetic differentiation. *Taxon* 60:721–732.

7. Comai L. 2005. The advantages and disadvantages of being polyploid. *Nature Reviews Genetics* 6:836–846.

8. Hojsgaard D, Hörandl E. 2015. A little bit of sex matters for genome evolution in asexual plants. *Frontiers in Plant Science* 6:82. https://doi.org/10.3389/fpls.2015.00082.

Page 25. Mixed Modes of Mating: St. John's wort (*Hypericum perforatum*)

1. Matzk F, Meister A, Brutovská R, Schubert I. 2001. Reconstruction of reproductive diversity in *Hypericum perforatum* L. opens novel strategies to manage apomixis. *The Plant Journal* 26:275–282.

2. Galla G, Barcaccia G, Schallau A, Molins MP, Bäumlein H, Sharbel TF. 2011. The cytohistological basis of apospory in *Hypericum perforatum* L. *Sexual Plant Reproduction* 24:47–61.

3. Mack RN, Erneberg M. 2002. The United States naturalized flora: Largely the product of deliberate introductions. *Annals of the Missouri Botanical Garden* 89:176–189.

4. Maron JL, Vilà M, Bommarco R, Elmendorf S, Beardsley P. 2004. Rapid evolution of an invasive plant. *Ecological Monographs* 74:261–280.

Page 26. Mating Types: yellow toadflax (*Linaria vulgaris*)

1. Bakshi TS, Coupland RT. 1960. Vegetative propagation in *Linaria vulgaris*. *Canadian Journal of Botany* 38:243–249.

2. Docherty Z. 1982. Self-incompatibility in *Linaria*. *Heredity* 49:349–352.

3. Bartlewicz J, Vandepitte K, Jacquemyn H, Honnay O. 2015. Population genetic diversity of the clonal self-incompatible herbaceous plant *Linaria vulgaris* along an urbanization gradient. *Biological Journal of the Linnean Society* 116:603–613.

4. Longley AA. 2019. *The Effect of Urbanization on Reproduction and Selection on Floral Traits in the Wildflower,* Linaria vulgaris. Master's Thesis. Guelph, Ontario: University of Guelph.

5. Desaegher J, Nadot S, Machon N, Colas B. 2019. How does urbanization affect the reproductive characteristics and ecological affinities of street plant communities? *Ecology and Evolution* 9:9977–9989.

Page 27. Genetic Mixing: white clover (*Trifolium repens*)

1. Burdon JJ. 1983. *Trifolium repens* L. *Journal of Ecology* 71:307–330.

2. Johnson MTJ, Prashad CM, Lavoignat M, Saini HS. 2018. Contrasting the effects of natural selection, genetic drift and gene flow on urban evolution in white clover (*Trifolium repens*). *Proceedings of the Royal Society B* 285:20181019. https://doi.org/10.1098/rspb.2018.1019.

3. Verboven HAF, Aertsen W, Brys R, Hermy M. 2014. Pollination and seed set of an obligatory outcrossing plant in an urban–peri-urban gradient. *Perspectives in Plant Ecology, Evolution and Systematics* 16:121–131.

4. Thompson L, Harper JL. 1988. The effect of grasses on the quality of transmitted radiation and its influence on the growth of white clover *Trifolium repens*. *Oecologia* 75:343–347.

5. Clifford HT. 1956. Seed dispersal on footwear. *Proceedings of the Botanical Society of the British Isles* 2:129–131; Schmidt W. 1989. Plant dispersal by motor cars. *Vegetatio* 80:147–152; Twigg LE, Lowe TJ, Taylor CM, Calver MC, Martin GR, Stevenson C, How R. 2009. The potential of seed-eating birds to spread viable seeds of weeds and other undesirable plants. *Austral Ecology* 34:805–820.

Page 28. Efficiency of Mating: common milkweed (*Asclepias syriaca*)

1. Brodschneider R, Crailsheim K. 2010. Nutrition and health in honey bees. *Apidologie* 41:278–294.

2. Wyatt R. 1976. Pollination and fruit-set in *Asclepias*: A reappraisal. *American Journal of Botany* 63:845–851.

3. Harder LD, Johnson SD. 2008. Function and evolution of aggregated pollen in angiosperms. *International Journal of Plant Sciences* 169:59–78.

4. Wyatt R, Broyles SB, Lipow SR. 2000. Pollen-ovule ratios in milkweeds (Asclepiadaceae): An exception that probes the rule. *Systematic Botany* 25:171–180.

5. Hargreaves AL, Harder LD, Johnson SD. 2009. Consumptive emasculation: The ecological and evolutionary consequences of pollen theft. *Biological Reviews* 84:259–276.

6. Frost SW. 1965. Insects and pollinia. *Ecology* 46:556–558.

Page 29. Exotic Hybrid Vigor: common reed (*Phragmites australis*)

1. Saltonstall K. 2002. Cryptic invasion by a non-native genotype of the common reed, *Phragmites australis*, into North America. *Proceedings of the National Academy of Sciences* 99:2445–2449.

2. Del Tredici P. 2010. *Wild Urban Plants of the Northeast: A Field Guide.* Ithaca, NY: Comstock Publishing Associates of Cornell University Press.

3. Lambert AM, Saltonstall K, Long R, Dudley TL. 2016. Biogeography of *Phragmites australis* lineages in the southwestern United States. *Biological Invasions* 18:2597–2617.

4. Williams J, Lambert AM, Long R, Saltonstall K. 2019. Does hybrid *Phragmites australis* differ from native and introduced lineages in reproductive, genetic, and morphological traits? *American Journal of*

Botany 106:29–41.

5. Saltonstall K, Lambert A, Rice N. 2016. What happens in Vegas, better stay in Vegas: *Phragmites australis* hybrids in the Las Vegas Wash. *Biological Invasions* 18:2463–2474.

Page 30. Variability in Fertility: mugwort (*Artemisia vulgaris*)

1. Del Tredici P. 2010. *Wild Urban Plants of the Northeast. A Field Guide.* Ithaca, NY: Comstock Publishing Associates of Cornell University Press.

2. Weston LA, Barney JN, DiTommaso A. 2005. A review of the biology and ecology of three invasive perennials in New York State: Japanese knotweed (*Polygonum cuspidatum*), mugwort (*Artemisia vulgaris*) and pale swallow-wort (*Vincetoxicum rossicum*). *Plant and Soil* 277:53–69.

3. Pawlowski F, Kapeluszny J, Kolasa A, Lecyk Z. 1967. Fertility of some species of ruderal weeds. *Annales Universitatis Mariae Curie-Sklodowska. Section E, Agriculture* 22:221–231; Dorph-Petersen K. 1925. Examinations of the occurrence and vitality of various weed species under different conditions, made at the Danish State Seed Testing Station during the years 1896–1923. In *Report of the Fourth International Seed Testing Congress,* 124–138. Cambridge, England: His Majesty's Stationery Office.

4. Martindale IC. 1876. The introduction of foreign plants. *Botanical Gazette* 2:55–58.

5. Barney JN, Whitlow TH, DiTommaso A. 2009. Evolution of an invasive phenotype: Shift to belowground dominance and enhanced competitive ability in the introduced range. *Plant Ecology* 202:275–284.

6. Barney JN. 2006. North American history of two invasive plant species: Phytogeographic distribution, dispersal vectors, and multiple introductions. *Biological Invasions* 8:703–717.

7. Garnock-Jones PJ. 1986. Floret specialization, seed production and gender in *Artemisia vulgaris* L. (Asteraceae, Anthemideae). *Botanical Journal of the Linnean Society* 92:285–302.

8. Uva RH, Neal JC, DiTomaso JM. 1997. *Weeds of the Northeast.* Ithaca, NY: Cornell University Press.

9. Onen H. 2007. Autotoxic potential of mugwort (*Artemisia vulgaris*). *Allelopathy Journal* 19:323–336.

Page 31. Hybrid Speciation: salsifies (*Tragopogon* spp.)

1. Ownbey M. 1950. Natural hybridization and amphiploidy in the genus *Tragopogon. American Journal of Botany* 37(7):487–499.

2. Soltis DE, Buggs RJA, Barbazuk WB, Schnable PS, Soltis PS. 2009. On the origins of species: Does evolution repeat itself in polyploid populations of independent origin? *Cold Spring Harbor Symposia on Quantitative Biology* 74:215–223.

3. Ownbey, Natural hybridization and amphiploidy in the genus *Tragopogon.*

4. Sinha RP, Whitehead FH. 1965. Meiotic studies of British populations of *Stellaria media* (L.) Vill., *S. neglecta* Weihe and *S. pallida* (Dumort.) Piré. *New Phytologist* 64:343–345; Paule J, Sharbel TF, Dobeš C. 2011. Apomictic and sexual lineages of the *Potentilla argentea* L. group (Rosaceae): Cytotype and molecular genetic differentiation. *Taxon* 60:721–732; Darmency H, Gasquez J. 1997. Spontaneous hybridization of the putative ancestors of the allotetraploid *Poa annua. New Phytologist* 136:497–501.

Page 32. Lost Sex: threadstalk speedwell (*Veronica filiformis*)

1. Mueller N, Sukopp H. 1993. Synanthropic distribution and association of the slender speedwell *Veronica filiformis* Smith. Tuexenia 13:399–413; Vinogradova YK, Kuklina AG, Galkina MA. 2017. The dynamics of clonal dispersal and regenerative activity of *Veronica filiformis* J.E. Smith. *Russian Journal of Biological Invasions* 8:197–205.

2. Scalone R, Albach D. 2014. Cytological evidence for gametophytic self-incompatibility in the genus *Veronica. Turkish Journal of Botany* 38:197–201.

3. Scalone R, Albach DC. 2012. Degradation of sexual reproduction in *Veronica filiformis* after introduction to Europe. *BMC Evolutionary Biology* 12:233. https://doi.org/10.1186/1471-2148-12-233; Uva RH, Neal JC, DiTomaso JM. 1997. *Weeds of the Northeast.* Ithaca, NY: Cornell University Press.

4. Šerá B. 2012. Which stem parts of slender speedwell (*Veronica filiformis*) are the most successful in plant regeneration? *Biologia* 67:110–115.

5. Harris GR, Lovell PH. 1980. Localized spread of *Veronica filiformis, V. agrestis and V. persica. The Journal of Applied Ecology* 17:815–826.

6. Scalone & Albach, Degradation of sexual reproduction in *Veronica filiformis* after introduction to Europe.

Page 33. Chapter 4 Trees and Woody Vines

1. Li H-L. 1963. *The Origin and Cultivation of Shade and Ornamental Trees.* Philadelphia: University of

Pennsylvania Press.

2. Rhoads AF, Block TA. 2005. *Trees of Pennsylvania: A Complete Reference Guide*. Philadelphia: University of Pennsylvania Press.

3. Johnson MG, Lang K, Manos P, Golet GH, Schierenbeck KA. 2016. Evidence for genetic erosion of a California native tree, *Platanus racemosa*, via recent, ongoing introgressive hybridization with an introduced ornamental species. *Conservation Genetics* 17:593–602.

Page 34. Segregation of Sexes: box elder (*Acer negundo*)

1. Overton RP. 1990. *Acer negundo*: L., Boxelder. In *Silvics of North America*, vol. 2, Hardwoods. Agriculture Handbook 654, edited by Burns RM, Honkala BH, 41–45. Washington, DC: Forest Service, United States Department of Agriculture.

2. Dawson T, Ehleringer JR. 1993. Gender-specific physiology, carbon isotope discrimination, and habitat distribution in boxelder, *Acer negundo*. *Ecology* 74:798–815.

3. Dawson and Ehleringer, Gender-specific physiology, 798–815.

4. Jing SW, Coley PD. 1990. Dioecy and herbivory: The effect of growth rate on plant defense in *Acer negundo*. *Oikos* 58:369–377.

5. Dawson and Ehleringer, Gender-specific physiology, 798-815.

Page 35. Sex Change in a Tree: silver maple (*Acer saccharinum*)

1. Gabriel WJ. 1990. *Acer saccharinum* L. Silver maple. In *Silvics of North America*, vol. 2, *Hardwoods*. *Agriculture Handbook 654*, edited by Burns RM, Honkala BH, 70–77. Washington, DC: Forest Service, United States Department of Agriculture.

2. Sakai AK, Oden NL. 1983. Spatial pattern of sex expression in silver maple (*Acer saccharinum* L.): Morisita's index and spatial autocorrelation. *American Naturalist* 122:489–508.

3. De Jong PC. 1976. *Flowering and Sex Expression in* Acer *L.: A Biosystematic Study*. Wageningen, the Netherlands: H. Veeman & Zonen B.V.

4. Sakai and Oden. Spatial pattern of sex expression in silver maple (*Acer saccharinum* L.)

5. Sakai and Oden. Spatial pattern of sex expression in silver maple (*Acer saccharinum* L.).

Page 36. Sex Change Over Days: Norway maple (*Acer platanoides*)

1. Nowak DJ, Rowntree RA. 1990. History and range of the Norway Maple. *Journal of Arboriculture* 16:291–296

2. Stout AB. 1938. The flowering behavior of Norway maples. *Journal of the New York Botanical Garden* 39:130–134.

3. Renner SS, Beenken L, Grimm GW, Kocyan A, Ricklefs RE. 2007. The evolution of dioecy, heterodichogamy, and labile sex expression in *Acer*. *Evolution* 61:2701–2719.

Page 37. Exotic Male Interference: bittersweets (*Celastrus* spp.)

1. Del Tredici P. 2014. Untangling the twisted tale of oriental bittersweet. *Arnoldia* 71:2–18.

2. Steward AM, Clemants SE, Moore G. 2003. The concurrent decline of the native *Celastrus scandens* and spread of the non-native *Celastrus orbiculatus* in the New York City metropolitan area. *The Journal of the Torrey Botanical Society* 130:143–146.

3. Zaya DN, Leicht-Young SA, Pavlovic NB, Feldheim KA, Ashley MV. 2015. Genetic characterization of hybridization between native and invasive bittersweet vines (*Celastrus* spp.). *Biological Invasions* 17:2975–2988.

Page 38. Exotic Male Dominance: mulberries (*Morus* spp.)

1. Klose N. 1963. Sericulture in the United States. *Agricultural History* 37:225–234.

2. Del Tredici P. 2010. *Wild Urban Plants of the Northeast: A Field Guide*. Ithaca, NY: Comstock Publishing Associates of Cornell University Press.

3. Clemants SE, Moore G. 2005. The changing flora of the New York metropolitan region. *Urban Habitats* 3:192–210.

4. Burgess KS, Husband BC. 2006. Habitat differentiation and the ecological costs of hybridization: The effects of introduced mulberry (*Morus alba*) on a native congener (*M. rubra*). *J. Ecol.* 94:1061–1069.

5. Burgess KS, Morgan M, Husband BC. 2008. Interspecific seed discounting and the fertility cost of hybridization in an endangered species. *New Phytologist* 177:276–283.

6. Burgess KS, Husband BC. 2006. Habitat differentiation and the ecological costs of hybridization: The effects of introduced mulberry (*Morus alba*) on a native congener (*M. rubra*). *Journal of Ecology* 94:1061–1069.

7. Taylor PE, Card G, House J, Dickinson MH, Flagan RC. 2006. High-speed pollen release in the white mulberry tree, *Morus alba* L. *Sexual Plant Reproduction* 19:19–24.

Page 40. Aphrodisiacs and Anti-aphrodisiacs: cabbage white (*Pieris rapae*)

1. Opler PA, Krizek GO. 1984. *Butterflies East of the Great Plains.* Baltimore: Johns Hopkins University Press.

2. Lia Y, Mathews RA. 2016. In vivo real-time monitoring of aphrodisiac pheromone release of small white cabbage butterflies (*Pieris rapae*). *Journal of Insect Physiology* 91–92:107–112.

3. Obara Y, Tateda H, Kuwabara M. 1975. Mating behavior of the cabbage white butterfly, *Pieris rapae crucivora* Boisduval. V.: Copulatory stimuli inducing changes of female response patterns. *Zoological Magazine* 84:71–76.

4. Andersson J, Borg-Karlson A-K, Wiklund C. 2000. Sexual cooperation and conflict in butterflies: A male-transferred anti-aphrodisiac reduces harassment of recently mated females. *Proceedings of the Royal Society B: Biological Sciences* 267:1271–1275.

5. Tigreros N, Sass EM, Lewis SM. 2013. Sex-specific response to nutrient limitation and its effects on female mating success in a gift-giving butterfly. *Evolutionary Ecology* 27:1145–1158; Tigreros N, Mowery MA, Lewis SM. 2014. Male mate choice favors more colorful females in the gift-giving cabbage butterfly. *Behavioral Ecology and Sociobiology* 68:1539–1547.

6. Stoehr AM, Hayes K, Wojan EM. 2016. Assessing the role of wing spots in intraspecific communication in the cabbage white butterfly (*Pieris rapae* L.) using a simple device to increase butterfly responses. *Journal of Insect Behavior* 29:243–255.

Page 41. Males Emerge First (Protandry): cabbage white (*Pieris rapae*)

1. Shapiro AM. 1970. The role of sexual behavior in density-related dispersal of pierid butterflies. *The American Naturalist* 104:367–372.

2. Obara Y. 1987. Avoidance of maladaptive, precocious copulation in the cabbage white butterfly, *Pieris rapae crucivora*. *Journal of Insect Physiology* 33:403–406.

3. Suzuki Y. 1979. Mating frequency in females of the small cabbage white, *Pieris rapae crucivora* Boisduval (Lepidoptera: Pieridae). *Kontyu* 47:335–339.

4. Ohsaki N. 1980. Comparative population studies of three Pieris butterflies, *P. rapae, P. melete* and *P. napi,* living in the same area: II. Utilization of patchy habitats by adults through migratory and non-migratory movements. *Research in Population Ecology* 22:163–183.

5. Jones RE, Gilbert N, Guppy M, Nealis V. 1980. Long-distance movement of *Pieris rapae*. *Journal of Animal Ecology* 49:629–642.

6. Wedell N, Cook PA. 1998. Determinants of paternity in a butterfly. *Proceedings: Biological Sciences* 265:625–630.

7. Takami Y, Koshio C, Ishii M, Fugii H, Hidaka T, Shimizu I. 2004. Genetic diversity and structure of urban populations of *Pieris* butterflies assessed using amplified fragment length. *Molecular Ecology* 13:245–258.

8. Rochat E, Manel S, Deschamps-Cottin M, Widmer I, Joost S. 2017. Persistence of butterfly populations in fragmented habitats along urban density gradients: Motility helps. *Heredity* 119:328–338.

Page 42. Average Is Best: male orange sulfur (*Colias eurytheme*)

1. Rutowski RL. 1985. Evidence for mate choice in a sulphur butterfly (*Colias eurytheme*). *Ethology* 70:103–114.

Page 43. Male Territoriality: red admiral butterfly (*Vanessa atalanta*)

1. Bitzer RJ, Shaw KC. 1979. Territorial behavior of the red admiral, *Vanessa atalanta* (L.) (Lepidoptera: Nymphalidae). *Journal of Research on the Lepidoptera* 18:36–49.

2. Takeuchi T. 2017. Agonistic display or courtship behavior? A review of contests over mating opportunity in butterflies. *Journal of Ethology* 35:3–12.

Page 44. Sex-Limited Mimicry: eastern tiger swallowtail (*Papilio glaucus*)

1. Brower JVZ. 1958. Experimental studies of mimicry in some North American butterflies. Part 2. *Battus*

philenor and *Papilio troilus*, *P. polyxenes*, and *P. glaucus*. *Evolution* 12:123–136; Jeffords MR, Sternburg JG, Waldbauer GP. 1979. Batesian mimicry: Field demonstration of the survival value of pipevine swallowtail and monarch color patterns. *Evolution* 33:275–286.

2. Krebs RA, West DA. 1988. Female mate preference and the evolution of female-limited Batesian mimicry. *Evolution* 42:1101–1104.

3. Ohsaki N. 1995. Preferential predation of female butterflies and the evolution of Batesian mimicry. *Nature* 378:173–175.

4. Scriber JM, Hagen RH, Lederhouse RC. 1996. Genetics of mimicry in the tiger swallowtail butterflies, *Papilio glaucus* and *P. canadensis* (Lepidoptera: Papilionidae). *Evolution* 50:222–236.

5. Brower LP, Brower JVZ. 1962. The relative abundance of model and mimic butterflies in natural populations of the *Battus philenor* mimicry complex. *Ecology* 1:154–158; Waldbauer GP, Sternburg JG. 1987. Experimental field demonstration that two aposematic butterfly color patterns do not confer protection against birds in northern Michigan. *The American Midland Naturalist* 118:145–152.

Page 46. Sequestered Females: evergreen bagworm moth (*Thyridopteryx ephemeraeformis*)

1. Kaufmann T. 1968. Observations on the biology and behavior of the evergreen bagworm moth, *Thyridopteryx ephemeraeformis* (Lepidoptera: Psychidae). *Annals of the Entomological Society of America* 61:38–44.

2. Kaufmann, Observations on the biology of the bagworm moth, 38–44.

3. Leonhardt BA, Neal JW Jr, Klun JA, Schwarz M, Plimmer JR. 1983. An unusual lepidopteran sex pheromone system in the bagworm moth. *Science* 219:314–316.

4. Jones F. 1927. The mating of Psychidae. *Transactions of the American Entomological Society* 53:293–313; plates 226–231.

5. Rhainds M, Leather S, Sadof C. 2008. Polyphagy, flightlessness, and reproductive output of females: A case study with bagworms (Lepidoptera: Psychidae). *Ecological Entomology* 33:663–672.

6. Haseman L. 1912. *The Evergreen Bagworm.* Columbia: University of Missouri College of Agriculture.

Page 47. Long-Distance Sex Pheromone: cecropia moth (*Hyalophora cecropia*)

1. Sternburg JG, Waldbauer GP, Scarbrough AG. 1981. Distribution of cecropia moth (Saturniidae) in central Illinois: A study in urban ecology. *Journal of the Lepidopterists' Society* 35:304–320; Mayer B. 2020. An urban population of saturniid silk moths. *News of the Lepidopterists' Society* 62:177–179.

2. Waldbauer GP, Sternburg JG. 1979. Inbreeding depression and a behavioral mechanism for its avoidance in *Hyalophora cecropia*. *The American Midland Naturalist* 102:204–208.

3. Waldbauer GP, Sternburg JG. 1982. Long mating flights by *Hyalophora cecropia* (L.) (Saturniidae). *Journal of the Lepidopterists' Society* 36:154–155.

4. Marsh FL. 1941. A few life-history details of *Samia cecropia* within the southwestern limits of Chicago. *Ecology* 22:331–337.

Page 48. Infertile Mating: ailanthus webworm moth (*Atteva aurea*)

1. Taylor ORJ. 1967. Relationship of multiple mating to fertility in *Atteva punctella* (Lepidoptera: Yponomeutidae). *Annals of the Entomological Society of America* 60:583–590.

Page 50. Matricide: bald-faced hornet (*Dolichovespula maculata*)

1. Bourke AF. 1994. Worker matricide in social bees and wasps. *Journal of Theoretical Biology* 167:283–292.

2. Bourke, Worker matricide, 283–292.

3. Loope KJ. 2015. Queen killing is linked to high worker-worker relatedness in a social wasp. *Current Biology* 25:2976–2979.

4. Akre RD, Myhre EA. 1992. Nesting biology and behavior of the baldfaced hornet, *Dolichovespula maculata* (L.) (Hymenoptera: Vespidae) in the Pacific Northwest. *Melanderia* 48:1–33; Foster KR et. al. 2001. Colony kin structure and male production in *Dolichovespula* wasps. *Mol. Ecol.* 10:1003–1010.

Page 51. Polyandry: common eastern bumblebee (*Bombus impatiens*)

1. Cnaani J, Schmid-Hempel R, Schmidt J. 2002. Colony development, larval development and worker reproduction in *Bombus impatiens* Cresson. *Insectes Sociaux* 49:164–170.

2. Baer B, Morgan ED, Schmid-Hempel P. 2001. A nonspecific fatty acid within the bumblebee mating plug prevents females from remating. *Proceedings of the National Academy of Sciences* 98:3926–3928.

3. Baer B, Schmid-Hempel P. 2001. Unexpected consequences of polyandry for parasitism and fitness in the bumblebee, *Bombus terrestris*. *Evolution* 55:1639–1643.

4. Cameron SA, Lozier JD, Strange JP, Koch JB, Cordes N, Solter LF, Griswold TL. 2011. Patterns of widespread decline in North American bumble bees. *Proceedings of the National Academy of Sciences* 108:662–667.

5. Owen RE, Whidden TL. 2013. Monandry and polyandry in three species of North American bumble bees (*Bombus*) determined using microsatellite DNA markers. *Canadian Journal of Zoology* 91:523–528.

6. Borge MA. 2016. Mysterious bumble bee behavior. *The Natural Web. Exploring Nature's Connections.* (Blog). https://the-natural-web.org/2016/11/13/mysterious-bumble-bee-behavior/.

7. Couvillon MJ, Fitzpatrick G, Dornhaus A. 2010. Ambient air temperature does not predict whether small or large workers forage in bumble bees (*Bombus impatiens*). *Psyche* 2010:536430. https://doi.org/536410.531155/532010/536430.

Page 52. Female Usurpers: common eastern bumblebee (*Bombus impatiens*)

1. Plath OE. 1922. Notes on the nesting habits of several North American bumblebees. *Psyche* 29:189–202; Plath OE. 1922. Notes on *Psithyrus* with records of two new American hosts. *Biological Bulletin of the Marine Biological Laboratory, Woods Hole* 43:23–44; Plath OE. 1923. Breeding experiments with confined *Bremus* (*Bombus*) queens. *Biological Bulletin* 45:325–341.

2. Plath, Notes on the nesting habits of several North American bumblebees.

3. Fisher RM. 1983. Behavioral interactions between a social parasite, *Psithyrus citrinus* (Hymenoptera: Apidae), and its bumble bee hosts. *Proceedings of the Entomological Society of Ontario* 114:55–60.

4. Fisher RM. 1984. Dominance by a bumble bee social parasite (*Psithyrus citrinus*) over workers of its host (*Bombus impatiens*). *Animal Behaviour* 32:304–305.

5. Zimma BO, Ayasse M, Tengö J, Ibarra F, Schulz C, Francke W. 2003. Do social parasitic bumblebees use chemical weapons? (Hymenoptera, Apidae). *Journal of Comparative Physiology A* 189:769–775.

6. Fisher RM. 1985. Evolution and host specificity: Dichotomous invasion success of *Psithyrus citrinus* (Hymenoptera: Apidae), a bumblebee social parasite in colonies of its two hosts. *Canadian Journal of Zoology* 63:977–981.

Page 53. Male Stuffing: European paper wasp (*Polistes dominula*)

1. Starks PT, Poe ES. 1997. 'Male-stuffing' in wasp societies. *Nature* 389:450.

Page 54. Leks: European paper wasp (*Polistes dominula*)

1. Höglund J, Alatalo RV. 1995. *Leks*. Princeton, NJ: Princeton University Press.

2. Beani L, Turillazzi S. 1990. Overlap at landmarks by lek-territorial and swarming males of two sympatric polistine wasps (Hymenoptera Vespidae). *Ethology Ecology & Evolution* 2:419–431.

3. Beani L, Turillazzi S. 1988. Alternative mating tactics in males of *Polistes dominulus* (Hymenoptera: Vespidae). *Behavioral Ecology and Sociobiology* 22:257–264.

4. Beani L, Zaccaroni M. 2015. Experimental male size manipulation in *Polistes dominula* paper wasps: Being the right size. *Ethology Ecology & Evolution* 27:185–199.

Page 55. Sons from Unfertilized Eggs: cicada killer (*Sphecius speciosus*)

1. Evans HE, O'Neill KM. 2007. *The Sand Wasps: Natural History and Behavior.* Cambridge, MA: Harvard University Press; Coelho JR. 1997. Sexual size dimorphism and flight behavior in cicada killers, *Sphecius speciosus*. *Oikos* 79:371–375.

2. Akre RD, Greene A, MacDonald JF, Landolt PJ, Davis HG. 1980. *The Yellowjackets of America North of Mexico. United States Department of Agriculture Handbook Number 552.* Washington, DC: United States Government Printing Office.

3. Dambach CA, Good E. 1943. Life history and habits of the cicada killer (*Sphecius speciosus*) in Ohio. *Ohio Journal of Science* 43:32–41.

4. de la Filia AG, Bain SA, Ross L. 2015. Haplodiploidy and the reproductive ecology of Arthropods. *Current Opinion in Insect Science* 9:36–43.

5. Fisher RA. 1930. *The Genetical Theory of Natural Selection.* London: Oxford University Press.

6. Grant PR. 2006. Opportunistic predation and offspring sex ratios of cicada-killer wasps (*Sphecius speciosus* Drury). *Ecological Entomology* 31:539–547.

Page 56. Lopsided Sex Ratios: pavement ant (*Tetramorium immigrans*)

1. Cordonnier M, Belleca A, Escarguel G, Kaufmann B. 2020. Effects of urbanization–climate interactions on range expansion in the invasive European pavement ant. *Basic and Applied Ecology* 44:46-54; Chin DWM, Bennett GW. 2018. Dominance of pavement ants (Hymenoptera: Formicidae) in residential areas of West Lafayette, IN, USA. *Journal of Entomological Science* 53:379–385.

2. Zhang YM, Vitone TR, Storer CG, Payton AC, Dunn RR, Hulcr J, McDaniel SF, Lucky A. 2019. From pavement to population genomics: characterizing a long-established non-native ant in North America through citizen science and ddRADseq. *Frontiers in Ecology and Evolution* 7:453. https://doi.org/10.3389/fevo.2019.00453.

3. Cordonnier M, Escarguel G, Dumet A, Kaufmann B. 2020. Multiple mating in the context of interspecific hybridization between two *Tetramorium* ant species. *Heredity* 124:675–684.

4.. Bruder KW, Gupta AP. 1972. Biology of the pavement ant, *Tetramorium caespitum* (Hymenoptera: Formicidae). *Annals of the Entomological Society of America* 65:358–367.

5. Brian MV, Elmes G, Kelly AF. 1967. Populations of the ant *Tetramorium caespitum* Latreille. *Journal of Animal Ecology* 36:337–342.

6. Passera L, Aron S, Vargo EL, Keller L. 2001. Queen control of sex ratio in fire ants. *Science* 293:1308–1310.

7. Kümmerli R, Keller L. 2009. Patterns of split sex ratio in ants have multiple evolutionary causes based on different within-colony conflicts. *Biological Letters* 5:713–716.

8. Sano K, Bannon N, Greene MJ. 2018. Pavement ant workers (*Tetramorium caespitum*) assess cues coded in cuticular hydrocarbons to recognize conspecific and heterospecific non-nestmate ants. *Journal of Insect Behavior* 31:186–199.

9 Zhang et al., From pavement to population genomics: characterizing a long-established non-native ant in North America through citizen science and ddRADseq.

Page 57. Chapter 7 Dragonflies and Praying Mantises

1. Pennsylvania Natural Heritage Program. 2008. Appendix VII: Odonates collected during Philadelphia County field surveys or other collections. In *A Natural Heritage Inventory of Philadelphia County, Pennsylvania, December 2008, Submitted to City of Philadelphia*, 173–174. Philadelphia, PA: Pennsylvania Natural Heritage Program.

Page 58. Male Territoriality: blue dasher dragonfly (*Pachydiplax longipennis*)

1. Olberg RM. 2011. Visual control of prey-capture flight in dragonflies. *Current opinion in Neurobiology* 22:267–271.

2. Sherk T. 1978. Development of the compound eyes of dragonflies (Odonata). III. Adult compound eyes. *Journal of Experimental Zoology* 203:61–68.

3. Futahashi R. 2016. Color vision and color formation in dragonflies. *Current Opinion in Insect Science* 17:32–39.

4. Kriska G, Bernáth B, Farkas R, Horváth G. 2009. Degrees of polarization of reflected light eliciting polarotaxis in dragonflies (Odonata), mayflies (Ephemeroptera) and tabanid flies (Tabanidae). *Journal of Insect Physiology* 55:1167–1173.

5. Stange G, Howard J. 1979. An ocellar dorsal light response in a dragonfly. *Journal of Experimental Biology* 83:351–355.

6. Robey CW. 1975. Observations on breeding behavior of *Pachydiplax longipennis* (Odonata: Libellulidae). *Psyche* 82:89–96.

7. Johnson C. 1962. A study of territoriality and breeding behavior in *Pachydiplax longipennis* Burmeister (Odonata: Libellulidae). *The Southwestern Naturalist* 7:191–197.

8. Robey, Observations on breeding behavior of *Pachydiplax longipennis* (Odonata: Libellulidae).

9. Sherman KJ. 1983. The adaptive significance of postcopulatory mate guarding in a dragonfly, *Pachydiplax longipennis*. *Animal Behaviour* 31:1107–1115.

Page 59. Secondary Genitalia: blue dasher dragonfly (*Pachydiplax longipennis*)

1. Johnson C. 1962. A study of territoriality and breeding behavior in *Pachydiplax longipennis* Burmeister (Odonata:Libellulidae). *The Southwestern Naturalist* 7:191–197; Robey CW. 1975. Observations on breeding behavior of *Pachydiplax longipennis* (Odonata: Libellulidae). *Psyche* 82:89–96; Utzeri C. 1985. Field observations on sperm translocation behaviour in the males of *Crocothemis erythraea* (Brullé) and

Orthetrum cancellatum (L.) (Libellulidae), with a review of the same in the Anisoptera. *Odonatologica* 14:227–237.

2. Sherman KJ. 1983. The adaptive significance of postcopulatory mate guarding in a dragonfly, *Pachydiplax longipennis. Animal Behaviour* 31:1107–1115.

Page 60. Female Mimicry of Males: blue dasher dragonfly (*Pachydiplax longipennis*)

1. Johnson C. 1962. A study of territoriality and breeding behavior in *Pachydiplax longipennis* Burmeister (Odonata:Libellulidae). *The Southwestern Naturalist* 7:191–197.

2. Berenbaum M. 2018. Damselflies in distress. *American Entomologist* 64:3–5; Gosden TP, Svensson EI. 2009. Density-dependent male mating harassment, female resistance, and male mimicry. *The American Naturalist* 173:709–721.

3. Dunkle SW. 1989. *Dragonflies of the Florida Peninsula, Bermuda and the Bahamas.* Gainesville, FL: Scientific Publishers.

4. Paulson D. 2011. *Dragonflies and Damselflies of the East: Princeton Field Guide.* Princeton, NJ: Princeton University Press.

Page 61. Polarized Light Pollution: dragonflies (**Odonata**)

1. Horváth G, Kriska G, Malik P, Robertson B. 2009. Polarized light pollution: A new kind of ecological photopollution. *Frontiers in Ecology and the Environment* 7:317–325.

2. Wildermuth H, Horváth G. 2005. Visual deception of a male *Libellula depressa* by the shiny surface of a parked car (Odonata: Libellulidae). *International Journal of Odonatology* 8:97–105.

3. Wildermuth H. 1998. Dragonflies recognize the water of rendezvous and oviposition sites by horizontally polarized light: A behavioural field test. *Naturwissenschaften* 85:297–302.

4. Rathod PP, Manwar NA, Raja IA. 2015. Visual deception in oviposition site selection in female dragonfly *Bradinopyga geminata* (Rambur) Libellulidae: Anisoptera. *International Journal of Advanced Research* 3:562–565.

5. Horváth G, Malik P, Kriska G, Wildermuth H. 2007. Ecological traps for dragonflies in a cemetery: The attraction of *Sympetrum* species (Odonata: Libellulidae) by horizontally polarizing black gravestones. *Freshwater Biology* 52:1700–1709.

6. Horváth G, Kriska G, Robertson B. 2014. Anthropogenic polarization and polarized light pollution inducing polarized ecological traps. In *Polarized Light and Polarization Vision in Animal Sciences*, edited by Horváth G, 443–513. Berlin-Heidelberg: Springer-Verlag.

Page 62. Monandry: fragile forktail damselfly (*Ischnura posita*)

1. Robinson JV, Allgeyer R. 1996. Covariation in life-history traits, demographics and behaviour in ischnuran damselflies: the evolution of monandry. *Biological Journal of the Linnean Society* 58:85-98.

2. Betros B. 2014. Photo of mating *Ischnura posita.* https://bugguide.net/node/view/1021772/bgimage; Anon. 2009. Photo of mating *Ischnura posita.* https://bugguide.net/node/view/278459/bgimage.

3. Robinson J. 1983. Effects of water mite parasitism on the demographics of an adult population of *Ischnura posita* (Hagen) (Odonata: Coenagrionidae). *American Midland Naturalist* 109:169-174.

Page 63. Sexual Cannibalism: praying mantises (**Family Mantidae**)

1. Hurd LE, Eisenberg R, Fagan W, Tilmon K, Snyder W, Vandersall K, Datz S, Welch J. 1994. Cannibalism reverses male-biased sex ratio in adult mantids: Female strategy against food limitation? *Oikos* 69:193–198.

2. Watanabe E, Adachi-Hagimori T, Miura K, Maxwell MR, Ando Y, Takematsu Y. 2011. Multiple paternity within field-collected egg cases of the praying mantid *Tenodera aridifolia. Annals of the Entomological Society of America* 104:348–352.

3. Roeder KD. 1935. An experimental analysis of the sexual behavior of the praying mantis (*Mantis religiosa* L.). *The Biological Bulletin* 69:203–220.

Page 64. Femme Fatale: praying mantises (**Family Mantidae**)

1. Kelner-Pillault S. 1957. Attirance sexuelle chez *Mantis religiosa* (Orth.). *Bulletin de la Société Entomologique de France* 62:9–11.

2. Barry KL. 2015. Sexual deception in a cannibalistic mating system? Testing the femme fatale hypothesis. *Proceedings of the Royal Society B* 282:0141428. https://doi.org/10.1098/rspb.2014.1428.

3. Barry KL, Holwell G, Herberstein M. 2008 Female praying mantids use sexual cannibalism as a foraging

strategy to increase fecundity. *Behavioral Ecology* 19:710–715.

4. Rau P, Rau N. 1913. Biology of *Stagmomantis carolina*. *Transactions of the Academy of Science of St. Louis* 22:1–58.

Page 65. Chapter 8 Other Insects

1. Jones A. 2020. New legislation targets bed bug infestations in Philadelphia. *Penn State College of Agriculture Department of Entomology News* (August 16, 2020). https://ento.psu.edu/news/2020/new-legislation-targets-bed-bug-infestations-in-philadelphia.

Page 66. Femme Fatale: fireflies (*Family Lampyridae*)

1. McDermott FA. 1911. Some further observations on the light-emission of American Lampyridae: The photogenic function as a mating adaptation in the Photinini. *Canadian Entomologist* 43:399–406.

2. Nelson S, Carlson AD, Copeland J. 1975. Mating-induced behavioural switch in female fireflies. *Nature* 255:628–629.

3. Lloyd JE. 1965. Aggressive mimicry in *Photuris*: firefly femmes fatales. *Science* 149:653–654.

4. Lloyd JE.1975. Aggressive mimicry in *Photuris* fireflies: Signal repertoires by femmes fatales. *Science* 187:452–453.

5. Faust LF. 2017. *Fireflies, Glow-worms, and Lightning Bugs*. Athens, Georgia: University of Georgia Press.

6. Owens ACS, Lewis SM. 2018. The impact of artificial light at night on nocturnal insects: A review and synthesis. *Ecology and Evolution* 2018:11337–11358.

Page 67. Contest Competition in Males: reddish-brown stag beetle (*Lucanus capreolus*)

1. Goyens J, Dirckx J, Aerts P. 2015. Stag beetle battle behavior and its associated anatomical adaptations. *Journal of Insect Behavior* 28:227–244.

2. Emlen DJ. 2014. Reproductive contests and the evolution of extreme weaponry. In *Evolution of Insect Mating System*s, edited by Shuker DM, Simmons LW, 92–105. Oxford: Oxford University Press.

3. Jeffords M. 2012. Beetle playground. In *Curious Encounters with the Natural World: From Grumpy Spiders to Hidden Tigers*, edited by Jeffords M, Post SL, 174–175. Champaign: University of Illinois Press.

4. Mathieu J. 1969. Mating behavior of five species of Lucanidae (Coleoptera: Insecta). *The Canadian Entomologist* 101:1054–1062.

Page 68. Vulnerable Meeting Place: locust borer (*Megacyllene robiniae*)

1. Galford JR. 1984. *The Locust Borer*. Forest Insect & Disease Leaflet 71. Washington, DC: Forest Service, U.S. Dept. of Agriculture.

2. Galford JR. 1977. *Evidence for a Pheromone in the Locust Borer*. Research Note NE-240. Upper Darby, PA: Northeastern Forest Experiment Station, Forest Service, U.S. Dept. of Agriculture.

3. Ginzel MD, Hanks LM. 2003. Contact pheromones as mate recognition cues of four species of longhorned beetles (Coleoptera: Cerambycidae). *Journal of Insect Behavior* 16:181–187; Ginzel MD, Millar JG, Hanks LM. 2003. (Z)-9-Pentacosene—contact sex pheromone of the locust borer, *Megacyllene robiniae*. *Chemical Ecology* 13:134–141.

4. Garman H. 1916. *The Locust Borer (Cyllene robiniae) and Other Insect Enemies of the Black Locust*. Bulletin No. 200. Lexington, KY: Kentucky Agricultural Experiment Station, State University; Harman DM, Harman AL. 1987. Distribution pattern of adult locust borers, (Coleoptera: Cerambycidae) on nearby goldenrod, *Solidago* spp. (Asteraceae), at a forest-field edge. *Proceedings of the Entomological Society of Washington* 89:706–710.

Page 69. Sexual Emigration: red milkweed beetle (*Tetraopes tetrophthalmus*)

1. Gontijo LM. 2013. Female beetles facilitate leaf feeding for males on toxic plants. *Ecological Entomology*:272–277.

2. Reagel PF, Ginzel MD, Hanks LM. 2002. Aggregation and mate location in the red milkweed beetle (Coleoptera: Cerambycidae). *Journal of Insect Behavior* 15:811–830.

3. Droney DC, Thaker M. 2006. Factors influencing mating duration and male choice in the red milkweed beetle, *Tetraopes tetrophthalmus* (Forster) (Coleoptera Cerambycidae). *Ethology Ecology & Evolution* 18:173–183.

4. Lawrence WS. 1986. Male choice and competition in *Tetraopes tetraophthalmus*: effects of local sex ratio variation. *Behavioral Ecology and Sociobiology* 18:289–296.

5. Lawrence WS. 1982. Sexual dimorphism in between and within patch movements of a monophagous

insect: *Tetraopes* (Coleoptera: Cerambycidae). *Oecologia* 53:245–250.

6. Lawrence WS. 1987. Dispersal: an alternative mating tactic conditional on sex ratio and body size. *Behavioral Ecology and Sociobiology* 21:367–373.

Page 70. Leks: northern house mosquito (*Culex pipiens*)

1. Harbach RE. 2012. *Culex pipiens*: species versus species complex – taxonomic history and perspective. *Journal of the American Mosquito Control Association* 28:10–23.

2. Knab F. 1906. The swarming of *Culex pipiens. Psyche: A Journal of Entomology* 13:123–134.

3. Tate P, Vincent M. 1936. The biology of autogenous and anautogenous races of *Culex pipiens* L. (Diptera). *Parasitology* 28:115–145.

4. Rozeboom LE, Gilford BN. 1954. Sexual isolation between populations of the *Culex pipiens* complex in *North America. Journal of Parasitology* 40:237–244; Spielman A. 2001. Structure and seasonality of Nearctic *pipiens* populations. *Annals of the New York Academy of Sciences* 951:220–234; Kilpatrick AM, Kramer LD, Jones MJ, Marra PP, Daszak P, Fonseca DM. 2007. Genetic influences on mosquito feeding behavior and the emergence of zoonotic pathogens. *American Journal of Tropical Medicine and Hygiene* 77:667–671; Kothera L, Godsey M, Mutebi J-P, Savage HM. 2010. A comparison of aboveground and belowground populations of *Culex pipiens* (Diptera: Culicidae) mosquitoes in Chicago, Illinois, and New York City, New York, using microsatellites. *Journal of Medical Entomology* 47:805–813.

5. Byrne K, Nichols RA. 1999. *Culex pipiens* in London Underground tunnels: Differentiation between surface and subterranean populations. *Heredity* 82:7–15; Kading RC. 2012 Studies on the origin of *Culex pipiens pipiens* form molestus in New York City. *Journal of the American Mosquito Control Association* 28:100–105.

6. Vinogradova EB. 2003. Ecophysiological and morphological variations in mosquitoes of the *Culex pipiens* complex (Diptera: Culicidae). *Acta Societatis Zoologicae Bohemicae* 67:41-50; Fritz ML, Walker ED, Miller JR, Severson DW, Dworkin I. 2015. Divergent host preferences of above- and below-ground *Culex pipiens* mosquitoes and their hybrid offspring. *Medical and Veterinary Entomology* 29:115–123.

7. Lindström A. 2017. History of human-biting *Culex pipiens* in Sweden and Scandinavia. *Journal of the European Mosquito Control Association* 35:10–12.

8. Osório C, Zé-Zé L, Amaro F, Nunes A, Alves MJ. 2014. Sympatric occurrence of *Culex pipiens* (Diptera, Culicidae) biotypes pipiens, molestus and their hybrids in Portugal, Western Europe: feeding patterns and habitat determinants. *Medical and Veterinary Entomology* 28:103–109.

Page 71. Facultative Parthenogenesis: oriental cockroach (*Blatta orientalis*)

1 Rehn JAG. 1945. Man's uninvited fellow traveler—the cockroach. *The Scientific Monthly* 61:265–276.

2 Dobney K, Kenward H, Ottaway P, Donel L. 1998. Down, but not out: Biological evidence for complex economic organization in Lincoln in the late fourth century. *Antiquity* 72:417–424.

3 Lihoreau M, Zimmer C, Rivaut C. 2007. Kin recognition and incest avoidance in a group-living insect. *Behavioral Ecology* 18:880–887; Lihoreau M, Rivault C. 2010. German cockroach males maximize their inclusive fitness by avoiding mating with kin. *Animal Behaviour* 20:303–309; Lihoreau M, Rivault C, Zweden JSV. 2016. Kin discrimination increases with odor distance in the German cockroach. *Behavioral Ecology* 27:1694–1701.

4 Abed D, Brossut R, Farine J-P. 1993. Evidence for sex pheromones produced by males and females in *Blatta orientalis* (Dictyoptera, Blattidae). *Journal of Chemical Ecology* 19:2831–2853.

5 Roth LM, Willis ER. 1956. Parthenogenisis in cockroaches. *Annals of the Entomological Society of America* 49:195–204.

6 Katoh K, Iwasaki M, Hosono S, Yoritsune A, Ochiai M, Mizunami M, Nishino H. 2017. Group-housed females promote production of asexual ootheca in American cockroaches. *Zoological Letters* 3:3. https://doi.org/10.1186/s40851-40017-40063-x.

7. Roth and Willis. Parthenogenesis in cockroaches.

Page 72. Parthenogenesis: milkweed aphid (*Aphis nerii*)

1. Harrison JS, Mondor EB. 2011. Evidence for an invasive aphid "superclone": Extremely low genetic diversity in Oleander aphid (*Aphis nerii*) populations in the southern United States. *PLOS ONE* 6:e17524. https://doi.org/17510.11371/journal.pone.0017524.

2. Blackman R. 1987. Reproduction, cytogenetics, and development. In *Aphids: Their Biology, Natural Enemies, and Control*, vol. 2A, edited by Minks AK, Harrewijn P, 163–195. Amsterdam: Elsevier.

Page 73. Cyclic Parthenogenesis: poplar leaf aphid (*Chaitophorus populicola*)

1. Richards WR. 1972. The Chaitophorinae of Canada (Homoptera: Aphididae). *Memoir of the Entomological Society of Canada* 104:1–109.

2. Palmer M. 1952. *Aphids of the Rocky Mountain region: including primarily Colorado and Utah, but also bordering area composed of southern Wyoming, southeastern Idaho and northern New Mexico.* Denver, CO: Thomas Say Foundation; Dixon AFG. 1987. Evolution and adaptive significance of cyclic parthenogenesis in aphids. In *Aphids: Their Biology, Natural Enemies and Control,* vol. 2A, edited by Minks AK, Harrewijn P, 289–297. Amsterdam: Elsevier.

3. Blackman R. 1987. Reproduction, cytogenetics, and development. In *Aphids: Their Biology, Natural Enemies, and Control,* vol. 2A, edited by Minks AK, Harrewijn P, 163–195. Amsterdam: Elsevier.

4. Simon J, Rispe C, Sunnucks P. 2002. Ecology and evolution of sex in aphids. *Trends in Ecology & Evolution* 17:34–39.

Page 74. Traumatic Insemination: bedbug (*Cimex lectularius*)

1. Siva-Jothy MT. 2006. Trauma, disease and collateral damage: Conflict in cimicids. *Philosophical Transactions of the Royal Society* B 361:269–275.

2. Reinhardt K, Naylor RA, Siva-Jothy MT. 2009. Situation exploitation: Higher male mating success when female resistance is reduced by feeding. *Evolution* 63:29–39.

3. Usinger RL. 1966. *Monograph of Cimicidae (Hemiptera-Heteroptera).* College Park, MD: Entomological Society of America.

4. Ryne C. 2009. Homosexual interactions in bed bugs: Alarm pheromones as male recognition signals. Animal Behaviour 78:1471–1475; Harraca V, Ryne C, Ignell R. 2010. Nymphs of the common bed bug (*Cimex lectularius*) produce anti-aphrodisiac defense against conspecific males. *BMC Biology* 8:121. https://doi.org/10.1186/1741-7007-8-121.

5. Stutt AD, Siva-Jothy MT. 2001. Traumatic insemination and sexual conflict in the bed bug *Cimex lectularius.* *Proceedings of the National Academy of Science* 98:5683–5687.

6. Reinhardt K, Naylor R, Siva-Jothy MT. 2003 Reducing a cost of traumatic insemination: Female bedbugs evolve a unique organ. *Proceedings of the Royal Society of London* B 270:2371–2375; Benoit JB, Jajack AJ, Yoder JA. 2012. Multiple traumatic insemination events reduce the ability of bed bug females to maintain water balance. *Journal of Comparative Physiology* B 182:189–198.

7. Morrow EH, Arnqvist G. 2003. Costly traumatic insemination and a female counter-adaptation in bed bugs. *Proceedings of the Royal Society of London* B 270:2377–2381; Reinhardt K, Naylor RA, Siva-Jothy MT. 2009. Ejaculate components delay reproductive senescence while elevating female reproductive rate in an insect. *Proceedings of the National Academy of Sciences* 106:21743–21747.

8. Pfiester M, Koehler PG, Pereira RM. 2009. Sexual conflict to the extreme: Traumatic insemination in bed bugs. *American Entomologist* 55:244–249.

9. Usinger. *Monograph of Cimicidae (Hemiptera- Heteroptera)*

10. Delaunay P. 2012. Human travel and traveling bedbugs. *Journal of Travel Medicine* 19:373–379.

11. Lange R, Reinhardt K, Michiels NK, Anthes N. 2013. Functions, diversity, and evolution of traumatic mating. *Biological Reviews* 88:585–601.

12. Tatarnic NJ, Cassis G, Siva-Jothy MT. 2014. Traumatic insemination in terrestrial arthropods. *Annual Review of Entomology* 59:245–261.

Page 75. Chapter 9 Invertebrates Exclusive of Insects

1. Chuang C-Y, Yang E-C, Tso I-M. 2008. Deceptive color signaling in the night: A nocturnal predator attracts prey with visual lures. *Behavioral Ecology and Sociobiology* 19:237–244; Peng P, Stuart-Fox D, Chen SW, Tan EJ, Kuo GL, Blamires SJ, Tso IM, Elgar MA. 2020. High contrast yellow mosaic patterns are prey attractants for orb-weaving spiders. *Functional Ecology* 34(4). https://doi.org/10.1111/1365-2435.13532; Ximenes NG, Moraes VDS, Ortega JCG, Gawryszewski FM. 2020. Color lures in orb-weaving spiders: A meta-analysis. *Behavioral Ecology* 31:568–576.

2. Schneider JM, Lubin Y. 1998. Intersexual conflict in spiders. *Oikos* 83:496–506.

3. Mouginot P, Prügel J, Thom U, Steinhoff POM, Kupryjanowicz J, Uhl G. 2015. Securing paternity by mutilating female genitalia in spiders. *Current Biology* 25:2980–2984.

Page 76. Sexual Mixing Around Outdoor Lights: bridge spider (*Larinioides sclopetarius*)

1. Levi HW. 1974. The orb-weaver genera *Araniella* and *Nuctenea* (Araneae: Araneidae). *Bulletin of the*

Museum of Comparative Zoology 146:291–316.

2. Kleinteich A. 2009. *Life History of the Bridge Spider,* Larinioides sclopetarius *(Clerck, 1757).* Dissertation. Hamburg: University of Hamburg.

3. Heiling A. 1999. Why do nocturnal orb-web spiders (Araneidae) search for light? *Behavioral Ecology and Sociobiology* 46:43–49.

4. Burgess JW, Uetz GW. 1982. Social spacing strategies in spiders. In *Spider Communication, Mechanisms and Ecological Significance*, edited by Witt PN, Rovner JS, 317–351. Princeton, NJ: Princeton University Press.

Page 77. Two Sets of Male Genitalia: bridge spider (*Larinioides sclopetarius*)

1. Eberhard WC, Huber BA. 2010. Spider genitalia: Precise maneuvers with a numb structure in a complex lock. In *The Evolution of Primary Sexual Characters in Animals*, edited by Leonard J, Cordoba-Aguilar A, 249–284. Oxford: Oxford University Press.

2. Bristowe WS. 1929. The mating habits of spiders, with special reference to the problems surrounding sex dimorphism. *Proceedings of the Zoological Society of London* 21:309–358.

3. Levi HW. 1974. The orb-weaver genera *Araniella* and *Nuctenea* (Araneae: Araneidae). *Bulletin of the Museum of Comparative Zoology* 146:291–316.

4. Montgomery TH Jr. 1903. Studies on the habits of spiders, particularly those of the mating period. *Proceedings of the Academy of Natural Sciences of Philadelphia* 55(Jan):59–149.

5. Dondale CD, Redner JH, Paquin P, Levi HW. 2003. *The Insects and Arachnids of Canada.* Part 23. The Orb-Weaving Spiders of Canada and Alaska. Araneae: Uloboridae, Tetragnathidae, Araneidae, Theridiosomatidae. Ottawa: National Research Council of Canada, National Research Council Press.

6. Eberhard et al. Spider genitalia.

Page 78. Sexually Compatible Personalities: bridge spider (*Larinioides sclopetarius*)

1. Finck A, Reed CF. 1979. Behavioral response to whole-body vibration in the orb-weaver *Araneus sericatus* Clerck (Araneae: Araneidae). *Journal of Arachnology* 7:139–147.

2. Vibert S, Scott C, Gries G. 2014. A meal or a male: The 'whispers' of black widow males do not trigger a predatory response in females. *Frontiers in Zoology* 11:4. https://doi.org/10.1186/1742-9994-1111-1184.

3. Montgomery TH Jr. 1903. Studies on the habits of spiders, particularly those of the mating period. *Proceedings of the Academy of Natural Sciences of Philadelphia* 55(Jan):59–149.

4. Gasket AC. 2007. Spider sex pheromones: Emission, reception, structures, and functions. *Biological Reviews* 82:27–48; Roland C. 1984. Chemical signals bound to the silk in spider communication (Arachnidae, Araneae). *The Journal of Arachnology* 11:309–314.

5. Levi HW. 1974. The orb-weaver genera *Araniella* and *Nuctenea* (Araneae: Araneidae). *Bulletin of the Museum of Comparative Zoology* 146:291–316.

6. Kralj-Fišer S, Mostajo GAS, Preik O, Pekár S, Schneider JM. 2013. Assortative mating by aggressiveness type in orb weaving spiders. *Behavioral Ecology* 24:824–831.

7. Kralj-Fišer S, Schneider JM. 2012. Individual behavioural consistency and plasticity in an urban spider. *Animal Behaviour* 84:197–204.

Page 79. Genetic Mixing in the City: house spiders (Araneae)

1. Johnson JC, Halpin R, Stevens D II, Vannan A, Lam J, Bratsch K. 2015. Individual variation in ballooning dispersal by black widow spiderlings: The effects of family and social rearing. *Current Zoology* 61:520–528.

2. Vetter RS, Flanders CP, Rust MK. 2009. The ability of spiderlings of the widow spider *Latrodectus hesperus* (Araneae: Theridiidae) to pass through different size mesh screen: Implications for exclusion from air intake ducts and greenhouses. *Journal of Economic Entomology* 102:1396–1398.

3. Choi MB, Lee SY, Yoo JS, Jun J, Kwon O. 2019. First record of the western black widow spider *Latrodectus hesperus* Chamberlin & Ivie, 1935 (Araneae: Theridiidae) in South Korea. *Entomological Research* 49:141–146; Jennings DT, McDaniel IN. 1988 *Latrodectus hesperus* (Araneae: Theridiidae) in Maine. *Entomological News* 99:37–40.

4. Miles LS, Dyer RJ, Verrelli BC. 2018. Urban hubs of connectivity: Contrasting patterns of gene flow within and among cities in the western black widow spider. *Proceedings of the Royal Society B* 285:1884; https://doi.org/10.1098/rspb.2018.1224; Miles LS, Johnson JC, Dyer RJ, Verrell BC. 2018. Urbanization as a facilitator of gene flow in a human health pest. *Molecular Ecology* 27:3219–3230.

5. Bayer S. 2014. Miscellaneous notes on European and African *Cheiracanthium* species (Araneae: Miturgidae). *Arachnologische Mitteilungen* 47:19–34; Denis J. 1936. On a collection of spiders from Algeria. *Journal of Zoology* 106:1027–1060; Guarisco H. 1999. House spiders of Kansas. *Journal of Arachnology* 27:217–221; Kazim M, Perveen R, Hussain R, Fatima N. 2014. Biodiversity of spiders (Arachnida: Araneae) of Karachi (Urbon) Sindh Province. *Pakistan Journal of Entomology and Zoology Studies* 2:308–313.

6. Vetter RS, Isbister GK, Bush SP, Boutin LJ. 2006. Verified bites by yellow sac spiders (genus *Cheiracanthium*) in the United States and Australia: Where is the necrosis? *American Journal of Tropical Medicine and Hygiene* 74:1043–1048.

Page 80. Mating Without Copulation: house centipede (*Scutigera coleoptrata*)

1. Acosta CA. 2003. The house centipede (*Scutigera coleoptrata*; Chilopoda): controversy and contradiction. *Journal of the Kentucky Academy of Science* 64:1–5.

2. Lewis JGE. 1981. *The Biology of Centipedes*. Cambridge, UK: Cambridge University Press.

3. Proctor HC. 1998. Indirect sperm transfer in arthropods: behavioral and evolutionary trends. *Annual Review of Entomology* 43:153–174.

4. Gergits WF, Jaeger RG. 1990. Field observations of the behavior of the red-backed salamander (*Plethodon cinereus*): courtship and agonistic interactions. *Journal of Herpetology* 24:93–95.

Page 81. Androgenesis: Asian clam (*Corbicula fluminea*)

1. Kraemer LR, Swanson C, Galloway M, Kraemer R. 1986. Biological basis of behavior in *Corbicula fluminea*. II: Functional morphology of reproduction and development and review of evidence for self-fertilization. *American Malacological Bulletin Special Edition* No 2:193–202.

2. Pigneur L-M, Hedtke SM, Etoundi E, Doninck KV. 2012. Androgenesis: A review through the study of the selfish shellfish *Corbicula* spp. *Heredity* 108:581–591.

3. Dawkins R. 1989. *The Selfish Gene*, 2nd ed. Oxford: Oxford University Press.

4. McMahon RF. 1999. Invasive characteristics of the freshwater bivalve *Corbicula fluminea*. In *Nonindigenous Freshwater Organisms: Vectors, Biology, and Impact*, edited by Claudi R, Leach JH, 315–343. Washington, DC: Lewis Publishers.

5. Harvey A, Szpilczak I. 2010. Potential effects of an invasive clam species on water treatment plants in Philadelphia: A case study of a significant die-off of *Corbicula fluminea* in the Schuylkill River. *Pennsylvania Section American Water Works Annual Conference*. Lancaster, PA; Ilarri MI, Antunes C, Guilhermino L, Sousa R. 2011. Massive mortality of the Asian clam *Corbicula fluminea* in a highly invaded area. *Biological Invasions* 13:277–280; McDowell WG, McDowell WH, Byers JE. 2016. Mass mortality of a dominant invasive species in response to an extreme climate event: Implications for ecosystem function. *Limnology and Oceanography* 62:177–188; Werner S, Rothhaupt K-O. 2008. Mass mortality of the invasive bivalve *Corbicula fluminea* induced by a severe low-water event and associated low water temperatures. *Hydrobiologia* 613:143–150.

6. Williams CJ, McMahon RF. 1986. Power station entrainment of *Corbicula fluminea* (Müller) in relation to population dynamics, reproductive cycle and biotic and abiotic variables. *American Malacological Bulletin*, Special Edition No. 2. 99–111.

Page 82. Facultative Self-Fertilization: gray garden slug (*Deroceras reticulatum*)

1. Wareing DR. 1986. Directional trail following in *Deroceras reticulatum*. *Journal of Molluscan Studies* 62:256–258.

2. Nicholas J. 1984. *The Biology of Reproduction in Two British Pulmonate Slugs*. Dissertation. Bangor, Wales, UK: University College of North Wales; Quick HE. 1960. British slugs (Pulmonata; Testacellidae, Arionidae, Limacidae). *Bulletin of the British Museum (Natural History) Zoology* 6:103–226; Reise H. 2007. A review of mating behavior in slugs of the genus *Deroceras* (Pulmonata: Agriolimacidae). *American Malacological Bulletin* 23:137–156.

3. Howlett SA. 2005. *The Biology, Behaviour and Control of the Field Slug* Deroceras reticulatum *(Müller)*. Dissertation. Newcastle upon Tyne: The University of Newcastle upon Tyne.

Page 83. Sexual Conflict of Interest: nightcrawler earthworm (*Lumbricus terrestris*)

1. Grove A. 1925. On the reproductive processes of the earthworm, *Lumbricus terrestris*. *Quarterly Journal of Microscopical Science* 69:245–291.

2. Koene JM, Pförtner T, Michiels NK. 2005. Piercing the partner's skin influences sperm uptake in the earthworm *Lumbricus terrestris*. *Behavioral Ecology and Sociobiology* 59:243–249.

3. Koene JM, Sundermann G, Michiels NK. 2002. On the function of body piercing during copulation in earthworms. *Invertebrate Reproduction & Development* 41:35–40.

4. Butt KR, Nuutinen V. 1998. Reproduction of the earthworm *Lumbricus terrestris* Linné after the first mating. *Canadian Journal of Zoology* 76:104–109.

5. Koene et al. On the function of body piercing during copulation in earthworms.

6. Johnstone R, Keller L. 2000. How males can gain by harming their mates: Sexual conflict, seminal toxins, and the cost of mating. *American Naturalist* 156:368–377; Mouginot P, Prügel J, Thom U, Steinhoff POM, Kupryjanowicz J, Uhl G. 2015. Securing paternity by mutilating female genitalia in spiders. *Current Biology* 25:2980-2984.

Page 84. Lethal Mating Trails: terrestrial flatworms (*Bipalium* spp.)

1. Ducey PK, Noce S. 1998. Successful invasion of New York State by the terrestrial flatworm, *Bipalium adventitium*. *Northeastern Naturalist* 5:199–206; Ogren RE. 1985. The human factor in the spread of an exotic land planarian in Pennsylvania. *Proceedings of the Pennsylvania Academy of Science* 59:117–118.

2. Nuutinen V, Butt KR. 1997. The mating behaviour of the earthworm *Lumbricus terrestris* (Oligochaeta: Lumbricidae). *Journal of the Zoological Society of London* 242:783–798; Nuutinen V, Butt K. 2005 Homing ability widens the sphere of influence of the earthworm *Lumbricus terrestris* L. *Soil Biology and Biochemistry* 37:805-807; Wetzel A, Uchman A, Bromley RG. 2016. Underground miners come out to the surface—trails of earthworms. *Ichnos* 23:97-107.

3. Fiore C, Tull JL, Zehner S, Ducey PK. 2004. Tracking and predation on earthworms by the invasive terrestrial planarian *Bipalium adventitium* (Tricladida, Platyhelminthes). *Behavioural Processes* 67:327–334.

4. Blackshaw RP. 1995. Changes in populations of the predatory flatworm *Artioposthia triangulata* and its earthworm prey in grassland. *Acta Zoologica Fennica* 196:107–110.

5. Keller E, Görres J, Schall J. 2017. Genetic structure of two invasive earthworms, *Amynthas agrestis* and *Amynthas tokioensis* (Oligochaeta, Megascolecidae), and a molecular method for species identification. *Megadrilogica* 22:140–149.

6. Gorsuch JP, Owen PC. 2014. Potential edaphic and aquatic predators of a nonindigenous Asian earthworm (*Amynthas agrestis*) in the eastern United States. *Northeastern Naturalist* 21:652–661.

Page 85. Chapter 10 Reptiles, Amphibians and Fish

1. Vargas-Salinas F, Cunnington GM, Amézquita A, Fahrig L. 2014. Does traffic noise alter calling time in frogs and toads? A case study of anurans in eastern Ontario, Canada. *Urban Ecosystems* 17:945–953.

Page 86. Anti-aphrodisiac Pheromone: common garter snake (*Thamnophis sirtalis*)

1. Shine R, Phillips B, Waye H, LeMaster M, Mason RT. 2003. Chemosensory cues allow courting male garter snakes to assess body length and body condition of potential mates. *Behavioral Ecology and Sociobiology* 54:162–166.

2. Lemaster MP, Moore IT, Mason RT. 2001. Conspecific trailing behaviour of red-sided garter snakes, *Thamnophis sirtalis parietalis*, in the natural environment. *Animal Behaviour* 61:827–833.

3. Mason RT, Jones TH, Fales HM, Pannell LK, Crews D. 1990. Characterization, synthesis, and behavioral responses to sex attractiveness pheromones of red-sided garter snakes (*Thamnophis sirtalis parietalis*). *Journal of Chemical Ecology* 16:2353–2369.

4. Halpert AP, Garstka WR, Crews D. 1982. Sperm transport and storage and its relation to the annual sexual cycle of the female red-sided garter snake, *Thamnophis sirtalis parietalis*. *Journal of Morphology* 174:149–159.

5. Carpenter CC. 1955. The garter snake. *The Scientific Monthly* 81:248–252.

6. Shine R, Olsson M, Mason RT. 2000. Chastity belts in gartersnakes: The functional significance of mating plugs. *Biological Journal of the Linnaean Society* 70:377–390.

7. Shine R, Mason RT. 2012. An airborne sex pheromone in snakes. *Biology Letters* 8:183–185.

8. Schwartz JM, McCracken GF, Burghardt GM. 1989. Multiple paternity in wild populations of the garter snake, *Thamnophis sirtalis*. *Behavioral Ecology and Sociobiology* 25:269–273; McCracken GF, Burghardt GM, Houts S. 1999. Microsatellite markers and multiple paternity in the garter snake *Thamnophis sirtalis*. *Molecular Ecology* 8:1475–1479.

9. Shine R, Lemaster MP, Wall M, Langkilde T, Mason RT. 2004. Why did the snake cross the road? Effects of roads on movement and location of mates by garter snakes (*Thamnophis sirtalis parietalis*). *Ecology and Society* 9(1). https://doi.org/10.5751/ES-00624-090109.

10. Burger J. 2001. The behavioral response of basking northern water (*Nerodia sipedon*) and eastern garter (*Thamnophis sirtalis*) snakes to pedestrians in a New Jersey park. *Urban Ecosystems* 5:19–129.

11. Gangloff EJ, Reding DM, Bertolatus D, Reigel CJ, Gagliardi-Seeley JL, Bronikowski AM. 2017. Snakes in the city: Population structure of sympatric gartersnakes (*Thamnophis* spp.) in an urban landscape. *Herpetological Conservation and Biology* 12:509–521.

Page 87. Mimetic Sex Pheromone: common garter snake (*Thamnophis sirtalis*)

1. LeMaster MP, Stefani A, Shine R, Mason RT. 2008. Cross-dressing in chemical cues: Exploring 'she-maleness' in newly-emerged male garter snakes. In *Chemical Signals in Vertebrates 11*, edited by Hurst JL, Beynon RJ, Roberts SC, Wyatt TD, 223–230. New York: Springer.

2. Shine R, Langkilde T, Mason RT. 2012. Facultative pheromonal mimicry in snakes: "She-males" attract courtship only when it is useful. *Behavioral Ecology and Sociobiology* 66:691–695.

3. Shine et al., Facultative pheromonal mimicry in snakes, 691–695..

4. Sexton OJ, Bramble JE. 1994. Post-hibernation behavior of a population of garter snakes (*Thamnophis sirtalis*). *Amphibia-Reptilia* 15:9–20.

5. Kubie JL. 1978. Roles of the vomeronasal and olfactory systems in courtship behavior of male garter snakes. *Journal of Comparative and Physiological Psychology* 92:627–641.

Page 88. Sex Determination by Temperature: red-eared slider turtle (*Trachemys scripta elegans*)

1. Rhodin AGJ, Iverson JB, Bour R, Fritz U, Georges A, Shaffer HB, van Dijk PP. 2017. *Turtles of the World: Annotated Checklist and Atlas of Taxonomy, Synonymy, Distribution, and Conservation Status*, 8th ed. Lunenburg, MA: Chelonian Research Foundation and Turtle Conservancy. https://doi.org/10.3854/crm.7.checklist.atlas.v8.2017.

2. Telecky TM. 2001. United States import and export of live turtles and tortoises. *Turtle and Tortoise Newsletter* 4:8–13.

3. Cagle ER. 1950. The life history of the slider turtle *Pseudemis scripta trootsii* (Holbrook). *Ecological Monographs* 20(1):31–54.

4. Cleiton F, Giuliano-Caetano L. 2008. Cytogenetic characterization of two turtle species: *Trachemys dorbigni* and *Trachemys scripta elegans*. *Caryologia* 61:253–257.

5. Bull JJ, Vogt RC, McCoy CJ. 1982. Sex determining temperatures in turtles: A geographic comparison. *Evolution* 36.326–332.

6. Ospina-Álvarez N, Piferrer F. 2008. Temperature-dependent sex determination in fish revisited: Prevalence, a single sex ratio response pattern, and possible effects of climate change. *PLOS ONE* 3:e2837. https://doi.org/10.1371/journal.pone.0002837; Pieau C, Dorizzi M, Richard-Mercier N. 1999. Temperature-dependent sex determination and gonadal differentiation in reptiles. *Cellular and Molecular Life Sciences* 55:887–900.

Page 89. Amplexus: American toad (*Anaxyrus americanus*)

1. Oldham RS. 1966. Spring movements in the American toad, *Bufo americanus*. *Canadian Journal of Zoology* 44:63–100.

2. Cunnington G, Fahrig L. 2010. Plasticity in the vocalizations of anurans in response to traffic noise. *Acta Oecologica* 36:463–470; Vargas-Salinas F, Cunnington GM, Amézquita A, Fahrig L. 2014. Does traffic noise alter calling time in frogs and toads? A case study of anurans in eastern Ontario, Canada. *Urban Ecosystems* 17:945–953.

3. Aronson LR. 1944. The sexual behavior of Anura. *American Museum Novitates* 1250:1–15.

4. Wright AH. 1914. *North American Anura: Life-Histories of the Anura of Ithaca, New York*. Washington, DC: Carnegie Institution of Washington.

5. Howard RD. 1988. Sexual selection on male body size and mating behaviour in American toads, *Bufo americanus*. *Animal Behaviour* 36:1796–1808.

6. Forester DC, Thompson KJ. 1998. Gauntlet behaviour as a male sexual tactic in the American toad (Amphibia: Bufonidae). *Behaviour* 135:99–119.

7. Howard RD, Young JR. 1998. Individual variation in male vocal traits and female mating preferences in *Bufo americanus*. *Animal Behaviour* 55:1165–1179.

8. Sullivan BK. 1992. Sexual selection and calling behavior in the American toad (*Bufo americanus*). *Copeia*

1992(1):1–7; Howard, Sexual selection on male body size and mating behaviour in American toads, *Bufo americanus*.

9. Waldman B, Rice JE, Honeycutt RL. 1992. Kin recognition and incest avoidance in toads. *Integrative and Comparative Biology* 32:18–30.

10. Price RM, Meyer ER. 1979. An amplexus call made by the male American toad, *Bufo americanus americanus* (Amphibia, Anura, Bufonidae). *Journal of Herpetology* 13:506–509.

11. Christein D, Taylor DH. 1978. Population dynamics in breeding aggregations of the American toad, *Bufo americanus* (Amphibia, Anura, Bufonidae). *Journal of Herpetology* 12:17–24.

12. Christein and Taylor, Population dynamics in breeding aggregations, 17–24.

13. Kruse KC, Mounce M. 1982. The effects of multiple matings on fertilization capability in male American toads (*Bufo americanus*). *Journal of Herpetology* 16:410–412; Gatz JJ. 1981. Non-random mating by size in American toads, *Bufo americanus*. *Animal Behaviour* 29:1004–1012.

Page 90. Alternative Mating Tactics: American toad (*Anaxyrus americanuss*)

1. Forester DC, Thompson KJ. 1998. Gauntlet behaviour as a male sexual tactic in the American toad (Amphibia: Bufonidae). *Behaviour* 135:99–119.

2. Kaminsky SK. 1997. *Bufo americanus* (American toad) reproduction. *Herpetological Review* 28:84. Forester, et al. Gauntlet behaviour as a male sexual tactic in the American toad.

3. Hitchings SP, Beebee TJC. 1998. Loss of genetic diversity and fitness in common toad (*Bufo bufo*) populations isolated by inimical habitat. *Journal of Evolutionary Biology* 11:269–283.

Page 91. Panmixia: American eel (*Anguilla rostrata*)

1. Cairns DK, Poirier LA, Murtojärvi M, Bernatchez L, Avery TS. 2017. *American Eel Distribution in Tidal Waters of The East Coast of North America, as Indicated by 26 Trawl and Beach Seine Surveys Between Labrador and Florida. Canadian Technical Report of Fisheries and Aquatic Sciences 3221.* Canada: Fisheries and Oceans Canada; Tesch F-W, Bartsch P, Berg R, Gabriel O, et al. 2003. *The Eel*, 3rd ed. Thorpe JE, Editor. Oxford: Blackwell.

2. Jessop BM. 1987. Migrating American eels in Nova Scotia. *Transactions of the American Fisheries Society* 116:161–170.

3. Munk P et al. 2010. Oceanic fronts in the Sargasso Sea control the early life and drift of Atlantic eels. *Proceedings of the Royal Society B* 277:3593–3599.

4. Schmidt J. 1923. The breeding places of the eel. *Philosophical Transactions of the Royal Society of London. Series B* 211:179–208.

5. Béguer-Pon M, Castonguay M, Shan S, Benchetrit J, Dodson JJ. 2015. Direct observations of American eels migrating across the continental shelf to the Sargasso Sea. *Nature Communications* 6:8705. https://doi.org/8710.1038/ncomms9705.

6. Pujolar JM. 2013. Conclusive evidence for panmixia in the American eel. *Molecular Ecology* 22:1761–1762.

7. Schabetsberger R, Miller MJ, Dall'Olmo G, Kaiser R, Økland F, Watanabe S, Aarestrup K, Tsukamoto K. 2016. Hydrographic features of anguillid spawning areas: Potential signposts for migrating eels. *Marine Ecological Progress Series* 554:141–155.

8. Aoyama J, Watanabe S, Miller MJ, Mochioka N, Otake T, Yoshinaga T, Tsukamoto K. 2014. Spawning sites of the Japanese eel in relation to oceanographic structure and the West Mariana Ridge. *PLOS ONE* 9:e88759. https://doi.org/88710.81371/journal.pone.0088759.

9. Shepard SL. 2015. *American Eel Biological Species Report. Supplement to: Endangered and Threatened Wildlife and Plants; 12-Month Petition Finding for the American Eel (*Anguilla rostrata*). Docket Number FWS-HQ-ES-2015-0143.* Hadley, MA: Fish and Wildlife Service, U.S. Dept. of Agriculture.

10. Limburg KE, Waldman JR. 2009. Dramatic declines in North Atlantic diadromous fishes. *BioScience* 59:955–965; Ezer T, Dangendorf S. 2020. Global sea level reconstruction for 1900–2015 reveals regional variability in ocean dynamics and an unprecedented long weakening in the Gulf Stream flow since the 1990s. *Ocean Science* 16:997–1016. https://doi.org/10.5194/os-16-997-2020; Kwak TJ, Engman AC, Lilyestrom CG. 2018. Ecology and conservation of the American eel in the Caribbean region. *Fisheries Management and Ecology* 26:42–52. https://doi.org/10.1111/fme.12300.

Page 92. Parenting Alone: male bluegill (*Lepomis macrochirus*)

1. Mecozzi M. 2008. Bluegill (*Lepomis macrochirus*). *PUBL-FM-711 08*. Madison, WI: Bureau of Fisheries

Management, Wisconsin Department of Natural Resources.

2. Miller HC. 1963. The behavior of the pumpkinseed sunfish *Lepomis gibbosus* (Linnaeus), with notes on the behavior of the other species of *Lepomis* and the pigmy sunfish, *Elassoma evergladei*. *Behaviour* 22:88–151.

3. Avila VL. 1976. A field study of nesting behavior of bluegill sunfish (*Lepomis macrochirus* Rafinesque). *The American Midland Naturalist* 96:195–206.

4. Coleman RM, Fischer RU. 1991. Brood size, male fanning effort and the energetics of a nonshareable parental investment in bluegill sunfish, *Lepomis macrochirus* (Teleostei: Centrarchidae). *Ethology* 87:177–188; Gross MR, Charnov EL. 1980. Alternative male life histories in bluegill sunfish. *Proceedings of the National Academy of Sciences* 77:6937–6940.

5. Neff BD. 2003. Decisions about parental care in response to perceived paternity. *Nature* 422:716–719.

6. Neff BD, Gross MR. 2001. Dynamic adjustment of parental care in response to perceived paternity. *Royal Society of London B: Biological Sciences* 268:1559–1565.

7. Neff BD. 2003. Paternity and condition affect cannibalistic behavior in nest-tending bluegill sunfish. *Behavioral Ecology and Sociobiology* 54:377–384.

8. Neff BD. 2001. Genetic paternity analysis and breeding success in bluegill sunfish (*Lepomis macrochirus*). *Journal of Heredity* 92:111–119.

Page 93. Sexual Mimicry: male bluegill (*Lepomis macrochirus*)

1. Gross MR, Charnov EL. 1980. Alternative male life histories in bluegill sunfish. *Proceedings of the National Academy of Sciences* 77:6937–6940.

2. Gross MR. 1982. Sneakers, satellites and parentals: Polymorphic mating strategies in North American sunfishes. *Zeitschrift für Tierpsychologie* 60:1–26.

3. Dominey WJ. 1980. Female mimicry in bluegill sunfish—A genetic polymorphism? *Nature* 284:546–548.

4. Gross. Sneakers, satellites and parentals.

5. Neff BD. 2004. Increased performance of offspring sired by parasitic males in bluegill sunfish. *Behavioral Ecology* 15:327–331.

6. Smith HM. 1907. *Fishes of North Carolina,* vol. II. Raleigh, NC: E. M. Uzzell & Co. plate 9. The caption reads: "Blue-gill; blue joe (*Lepomis incisor*). Female."

Page 94. Feminizing Pollution: bluegill (*Lepomis macrochirus*)

1. Lee Pow CS, Law JM, Kwak TJ, Cope WG, Rice JA, Kullman SW, Aday DD. 2017. Endocrine active contaminants in aquatic systems and intersex in common sport fishes. *Environmental Toxicology and Chemistry* 36:959–968.

Page 95. Chapter 11 Mammals

1. Egnor SR, Seagraves KM. 2016. The contribution of ultrasonic vocalizations to mouse courtship. *Current Opinion in Neurobiology* 38:1–5.

2. Marsden HM, Holler NR. 1964. Social behavior in confined populations of the cottontail and the swamp rabbit. *Wildlife Monographs* 13:3–39.

3. Lishak RS. 1982. Gray squirrel mating calls: A spectrographic and ontogenic analysis. *Journal of Mammalogy* 63:661–663; Koprowski JL. 1993. Do estrous female gray squirrels *Sciurus carolinensis* advertise their receptivity? *Canadian Field-Naturalist* 106:392–394; Thompson DC. 1977. Reproductive behavior of the grey squirrel. *Canadian Journal of Zoology* 55:1176–1184.

Page 96. Territoriality: eastern gray squirrel (*Sciurus carolinensis*)

1. Thompson DC. 1978. The social system of the grey squirrel. *Behaviour* 64:305–328.

2. Lawton C. 2016. Ranging behaviour, density and social structure in grey squirrels. In *Grey Squirrels: Ecology and Management of an Invasive Species in Europe*, edited by Shuttleworth CM, Lurz PWW, Gurnell J, 133–152. Suffolk, England: European Squirrel Initiative; Koprowski JL, Koprowski N. 1987. Joint nest-building activity in the eastern gray squirrel, *Sciurus carolinensis*. *Canadian Field-Naturalist* 101:610–611.

3. Koprowski JL. 1996. Natal philopatry, communal nesting, and kinship in fox squirrels and eastern gray squirrels. *Journal of Mammalogy* 77:1006–1016.

4. Hadidian J, Manski D, Flyger V, Cox C, Hodge G. 1987. Urban gray squirrel damage and population management: A case history. In *Proceedings of the Third Eastern Wildlife Damage Control Conference,*

edited by Holler NR, 219–227. Gulf Shores, AL: Eastern Wildlife Damage Control Conference.

Page 97. Mating Plug: female eastern gray squirrel (*Sciurus carolinensis*)

1. Baer B, Morgan ED, Schmid-Hempel P. 2001. A nonspecific fatty acid within the bumblebee mating plug prevents females from remating. *Proceedings of the National Academy of Sciences* 98:3926–3928; Friesen CR, Shine R, Krohmer RW, Mason RT. 2013. Not just a chastity belt: The functional significance of mating plugs in garter snakes, revisited. *Biological Journal of the Linnean Society* 109:893–907; Lewis SM, Cratsley CK, Rooney JA. 2004. Nuptial gifts and sexual selection in *Photinus* fireflies. *Integrative and Comparative Biology* 44:234–237; Watanabe M, Sato K. 1993. A spermatophore structured in the bursa copulatrix of the small white *Pieris rapae* (Lepidoptera, Pieridae) during copulation, and its sugar content. *Journal of Research on the Lepidoptera* 32:26–36.

2. Koprowski JL. 1992. Removal of copulatory plugs by female tree squirrels. *Journal of Mammalogy* 73:572–576.

3. Solomon NG, Keane B. 2007. Reproductive strategies in female rodents. In *Rodent Societies: An Ecological and Evolutionary Perspective*, edited by Wolff JO, Sherman PW, 42–56. Chicago: University of Chicago Press.

Page 98. Alternative Mating Tactic: male eastern gray squirrel (*Sciurus carolinensis*)

1. Koprowski JL. 1993. Alternative reproductive tactics in male eastern gray squirrels: "Making the best of a bad job." *Behavioral Ecology* 4:165–171.

Page 99. Copulation-Induced Ovulation: eastern cottontail (*Sylvilagus floridanus*)

1. Chapman JA, Hockman JG, Ojeda MM. 1980. *Sylvilagus floridanus*. *Mammalian Species* 136:1–8.; Baker MAA, Emerson SE, Brown JS. 2015. Foraging and habitat use of eastern cottontails (*Sylvilagus floridanus*) in an urban landscape. *Urban Ecosystems* 18:977–987.

2. Kolb HH. 1985. The burrow structure of the European rabbit (*Oryctolagus cuniculus* L.). *Journal of Zoology* 206:253–262.

3. Hunt VM, Magle SB, Vargas C, Brown AW, Lonsdorf EV, Sacerdote AB, Sorley EJ, Santymire RM. 2014. Survival, abundance, and capture rate of eastern cottontail rabbits in an urban park. *Urban Ecosystems* 17:547–560.

4. Conaway CH, Wight HM, Sadler KC. 1963. Annual production by a cottontail population. *The Journal of Wildlife Management* 27:171–175.

5. Wight HM, Conaway CH. 1962. Determination of pregnancy rates of cottontail rabbits. *The Journal of Wildlife Management* 26:93–95; Adams G, Ratto M, Silva M, Carrasco R. 2016. Ovulation-inducing factor (OIF/NGF) in seminal plasma: A review and update. *Reproduction in Domestic Animals* 51:4–17.

6. Casteel DA. 1967. Timing of ovulation and implantation in the cottontail rabbit. *The Journal of Wildlife Management* 31:194–197.

7. Pincus G, Enzmann EV. 1932. Fertilisation in the rabbit. *Journal of Experimental Biology* 9:403–408.

8. Staples RE. 1967. Behavioral induction of ovulation in the oestrous rabbit. *Journal of Reproduction and Fertility* 13:429–435.

9. Silva M, Niño A, Guerra M, Letelier C, Valderrama XP, Adams GP, Ratto MH. 2011. Is an ovulation-inducing factor (OIF) present in the seminal plasma of rabbits? *Animal Reproduction Science* 127(3-4):213–221.

Page 100. Pregnancy Blocker: house mouse (*Mus musculus*)

1. Berry RJ, Bronson FH. 1992. Life history and bioeconomy of the house mouse. *Biological Reviews* 67:519-550.

2. Whitten WK. 1956. Modification of the oestrous cycle of the mouse by external stimuli associated with the male. *Journal of Endocrinology* 13:399–404: Whitten W, Bronson F, Greenstein J. 1968. Estrus-inducing pheromone of male mice: transport by movement of air. *Science* 161 584–585.

3. Bruce H. 1959. Exteroceptive block to pregnancy in the mouse. *Nature* 184:105; Parkes A, Bruce HM. 1962. Pregnancy-block in female mice placed in boxes soiled by males. *Journal of Reproduction and Fertility* 4:303–308.

4. Eccard JA, Dammhahn M, Ylönen H. 2017. The Bruce effect revisited: Is pregnancy termination in female rodents an adaptation to ensure breeding success after male turnover in low densities? *Oecologia* 185:81–94.

5. Ferrari M, Lindholm A, König B. 2019. Fitness consequences of female alternative reproductive tactics in

house mice (*Mus musculus domesticus*). *American Naturalist* 193:106–124.

6. Chipman RK, Fox KA. 1966. Oestrous synchronization and pregnancy blocking in wild house mice (*Mus musculus*) in reproduction. *Journal of Reproduction and Fertility* 12:233–236.

7. Thonhauser KE, Thoß M, Musolf K, Klaus T, Penn DJ. 2014. Multiple paternity in wild house mice (*Mus musculus musculus*): effects on offspring genetic diversity and body mass. *Ecology and Evolution* 4:200–209.

Page 102. Territoriality: red-tailed hawk (*Buteo jamaicensis*)

1. Fitch HS, Swenson F, Tillotson DF. 1946. Behavior and food habits of the red-tailed hawk. *The Condor* 48:205–237.

2. Rothfels M, Lein MR. 1983. Territoriality in sympatric populations of red-tailed and Swainson's hawks. *Canadian Journal of Zoology* 61:60–64; Wiley JW. 1975. Three adult red-tailed hawks tending a nest. *The Condor* 77:480–482.

3. Ballam JM. 1984. The use of soaring by the red-tailed hawk (*Buteo jamaicensis*). *The Auk* 101:519–524.

4. Janes SW. 1984. Fidelity to breeding territory in a population of red-tailed hawks. *The Condor* 86:200–203.

5. Brown B. 2018. Hawks and hawkaholics (Blog). https://www.gridphilly.com/blog-home/2018/4/11/a-community-of-admirers-advocates-for-the-red-tailed-hawk; Hunold C. 2017. Why not the city? Urban hawk watching and the end of nature. *Nature and Culture* 12:115–136.

6. Preston CR. 2000. *Wild Bird Guides: Red-Tailed Hawk.* Mechanicsburg, PA: Stackpole.

Page 103. Conditional Sexual Trait: house finch (*Haemorhous mexicanus*)

1. Hill GE. 1990. Female house finches prefer colourful males: Sexual selection for a condition-dependent trait. *Animal Behaviour* 40:563–572.

2. Hill GE, Nolan PM, Stoehr AM. 1999. Pairing success relative to male plumage redness and pigment symmetry in the house finch: Temporal and geographic constancy. *Behavioral Ecology* 10:48–53; McGraw KJ, Stoehr AM, Nolan PM, Hill GE. 2001. Plumage redness predicts breeding onset and reproductive success in the house finch: A validation of Darwin's theory. *Journal of Avian Biology* 32:90–94.

3. Hill GE. 1991. Plumage coloration is a sexually selected indicator of male quality. *Nature* 350:337–339.

4. Hill GE. 1992. The proximate basis of variation in carotenoid pigmentation in male house finches. *The Auk* 109:1–2.

5. McGraw KJ, Nolan PM, Crino OL. 2011. Carotenoids bolster immunity during moult in a wild songbird with sexually selected plumage coloration. *Biological Journal of the Linnean Society* 102:560–572.

6. Hill GE, Farmer KL. 2005. Carotenoid-based plumage coloration predicts resistance to a novel parasite in the house finch. *Naturwissenschaften* 92(1):30–34; Hochachka WM, Dhondt AA. 2000. Density-dependent decline of host abundance resulting from a new infectious disease. *Proceedings of the National Academy of Science* 97(10):5303–5306.

7. Badyaev AV, Hill GE, Dunn PO, Glen JC. 2001. Plumage color as a composite trait: Developmental and functional integration of sexual ornamentation. *The American Naturalist* 158:221–235; Hill, Plumage coloration is a sexually selected indicator of male quality.

8. Duckworth RA, Mendonça MT, Hill GE. 2004. Condition-dependent sexual traits and social dominance in the house finch. *Behavioral Ecology* 15:779–784.

9. Darwin C. 1871. *On the Origin of Species,* 5th ed. New York: Appleton and Co.

10. Hill GE. 1994. Geographic variation in male ornamentation and female mate preference in the house finch: A comparative test of models of sexual selection. *Behavioral Ecology* 5:64–73.

11. Prum RO. 2017. *The Evolution of Beauty.* New York: Doubleday.

Page 104. Submissive Male: house finch (*Haemorhous mexicanus*)

1. Thompson WL. 1960. Agonistic behavior in the house finch. Part I: Annual cycle and display patterns. *The Condor* 62:245–271; McGraw KJ, Hill GE. 2000. Plumage brightness and breeding-season dominance in the house finch: A negatively correlated handicap? *The Condor* 102:456–461.

2. McGraw KJ, Medina-Jerez W, Adams H. 2007. Carotenoid-based plumage coloration and aggression during molt in male house finches. *Behaviour* 144:165–178.; McGraw KJ, Hill GE. 2000. Carotenoid-based ornamentation and status signaling in the house finch. *Behavioral Ecology* 11:520–527.

3. Hasegawa M, Ligon RA, Giraudeau M, Watanabe M, McGraw KJ. 2014. Urban and colorful male house

finches are less aggressive. *Behavioral Ecology* 25:641–649.

4. Badyaev AV, Young RL, Oh KP, Addison C. 2008. Evolution on a local scale: Developmental, functional, and genetic bases of divergence in bill form and associated changes in song structure between adjacent habitats. *Evolution* 62:1951–1964.

5. Hasegawa, Ligon, Giraudeau, Watanabe, McGraw. Urban and colorful male house finches are less aggressive.

Page 105. Courtship Feeding: house finch (*Haemorhous mexicanus*)

1. Anderson AH, Anderson A. 1944. 'Courtship feeding' by the house finch. *The Auk* 61:477–478.

2. Lack D. 1940. Courtship feeding in birds. *The Auk* 57:169–178.

3. Galván I, Sanz JJ. 2011. Mate-feeding has evolved as a compensatory energetic strategy that affects breeding success in birds. *Behavioral Ecology* 22:1088–1095.

4. Hill GE. 2002. *A Red Bird in a Brown Bag: The Function and Evolution of Colorful Plumage in the House Finch*. New York: Oxford University Press.

Page 106. Repeated Broods: mourning dove (*Zenaida macroura*)

1. Otis DL, Schulz JH, Miller D, Mirarchi RE, Baskett TS. 2008. Mourning dove (*Zenaida macroura*), version 2.0. In *The Birds of North America*, edited by Rodewald PG . Ithaca, NY: Cornell Laboratory of Ornithology. https://doi.org/10.2173/bna.117.

2. Leopold AS, Dedon MF. 1983. Resident mourning doves in Berkeley, California. *The Journal of Wildlife Management* 47:780–789.

3. Muñoz AM, McCleery RA, Lopez RR, Silvy NJ. 2008. Nesting ecology of mourning doves in an urban landscape. *Urban Ecosystems* 11:257–267.

4. Westmoreland D, Best LB, Blockstein DE. 1986. Multiple brooding as a reproductive strategy: Time-conserving adaptations in mourning doves. *The Auk* 103(1):196–203.

5. Böhning-Gaese K, Halbe B, Lemoine N, Oberrath R. 2000. Factors influencing the clutch size, number of broods and annual fecundity of North American and European land birds. *Evolutionary Ecology Research* 2:823–839.

6. Blockstein DE, Westmoreland D. 1993. Reproductive strategy. In *Ecology and Management of the Mourning Dove*, edited by Baskett TS, Sayre MW, Tomlinson RE, Mirarchi RE, 105–116. Harrisburg, PA: Stackpole Books.

7. Cheng MF, Porter M, Ball G. 1981. Do ring doves copulate more than necessary for fertilization? *Physiology & Behavior* 27:659–662.

8. Petrie M. 1992. Copulation frequency in birds: Why do females copulate more than once with the same male? *Animal Behaviour* 44:790–792.

9. Møller AP. 1987. Copulation behaviour in the goshawk, *Accipiter gentilis*. *Animal Behaviour* 35:755–763.

Page 107. Shortened Courtship: mourning dove (*Zenaida macroura*)

1. Tori GM, Peterle TJ. 1983. Effects of PCBs on mourning dove courtship behavior. *Bulletin of Environmental Contamination and Toxicology* 30:44–49.

2. Sayre MW, Baskett TS, Mirarchi RE. 1993. Behavior. *In Ecology and Management of the Mourning Dove*, edited by Baskett TS, Sayre MW, Tomlinson RE, Mirarchi RE, 161–180. Harrisburg, PA: Stackpole Books; Jackson GL, Baskett TS. 1964. Perch-cooing and other aspects of breeding behavior of mourning doves. T*he Journal of Wildlife Management* 28:293–307.

3. Westmoreland D, Best LB, Blockstein DE. 1986. Multiple brooding as a reproductive strategy: Time-conserving adaptations in mourning doves. *The Auk* 103:196–203.

Page 111. Coping with Noise: mourning dove (*Zenaida macroura*)

1. Francis CD, Ortega CP, Cruz A. 2009. Noise pollution changes avian communities and species interactions. *Current Biology* 19:1415–1419; Francis CD, Ortega CP, Cruz A. 2011. Noise pollution filters bird communities based on vocal frequency. *PLOS ONE* 6:e27052. https://doi.org/10.5061/dryad.75nn1932.

2. Bermúdez-Cuamatzin E, Ríos-Chelén AA, Gil D, Garcia CM. 2011. Experimental evidence for real-time song frequency shift in response to urban noise in a passerine bird. *Biology Letters* 7:36–38; Guo F, Bonebrake TC, Dingle C. 2016. Low frequency dove coos vary across noise gradients in an urbanized environment. *Behavioral Processes* 129:86–93.

3. Cheng MF, Peng JP, Johnson P. 1998. Hypothalamic neurons preferentially respond to female nest coo

stimulation: Demonstration of direct acoustic stimulation of luteinizing hormone release. *The Journal of Neuroscience* 18:5477–5489.

4. Halfwerk W, Holleman LJM, Lessells CKM, Slabbekoorn H. 2011. Negative impact of traffic noise on avian reproductive success. *Journal of Applied Ecology* 48:210–219.

5. Holt DE, Johnston CE. 2015. Traffic noise masks acoustic signals of freshwater stream fish. *Biological Conservation* 187:27–33.

6. de Jong K, Amorim MCP, Fonseca PJ, Fox CJ, Heubel KU. 2018. Noise can affect acoustic communication and subsequent spawning success in fish. *Environmental Pollution* 237:814–823.

7. Lampe U, Schmoll T, Franzke A, Reinhold K. 2012. Staying tuned: Grasshoppers from noisy roadside habitats produce courtship signals with elevated frequency components. *Functional Ecology* 26:1348–1354; Orci KM, Petroczki K, Barta Z. 2016. Instantaneous song modification in response to fluctuating traffic noise in the tree cricket *Oecanthus pellucens*. *Animal Behavior* 112:187–194.

8. Wu C-H, Elias DO. 2014. Vibratory noise in anthropogenic habitats and its effect on prey detection in a web-building spider. *Animal Behaviour* 90:47–56.

9. Halfwerk W, Lohr B, Slabbekoorn H. 2018. Impact of man-made sound on birds and their songs. In *Effects of Anthropogenic Noise on Animals*, edited by Slabbekoorn H, Dooling R, Popper AN, Fay RR, 209–242. New York: Springer.

Page 112. Sexual Singing: American robin (*Turdus migratorius*)

1. Wauer RH. 1999. *The American Robin*. Austin, TX: University of Texas Press.

2. Seger-Fullam KD, Rodewald AD, Soha JA. 2011. Urban noise predicts song frequency in northern cardinals and American robins. *Bioacoustics: The International Journal of Animal Sound and Its Recording* 20:267-276; Dowling JL, Luther DA, Marra PP. 2012. Comparative effects of urban development and anthropogenic noise on bird songs. *Behavioral Ecology* 23:201-209; Miller MW. 2006. Apparent effects of light pollution on singing behavior of American robins. *The Condor* 108:130–139.

3. Johnson SL. 2006. Do American robins acquire songs by both imitating and inventing? *The Wilson Journal of Ornithology* 118:341–352;

4. Slagsvold T. 1996. Dawn and dusk singing of male American robins in relation to female behavior. *The Wilson Bulletin* 108:507–515.

5. Morneau F, Lépinea C, Décariea R, Villardb M-A, DesGrangesc J-L. 1995. Reproduction of American robin (*Turdus migratorius*) in a suburban environment. *Landscape and Urban Planning* 32:55–62.

Page 113. Four Sexes: white-throated sparrow (*Zonotrichia albicollis*)

1. Lowther JK. 1961. Polymorphism in the white-throated sparrow, *Zonotrichia albicollis* (Gmelin). *Canadian Journal of Zoology* 39:281–292.

2. Lowther JK, Falls JB. 1968. *Zonotrichia albicollis* (Gmelin): White-throated sparrow. In *Life Histories of North American Cardinals, Grosbeaks, Buntings, Towhees, Finches, Sparrows, and Allies*. United States National Museum Bulletin 237, edited by Bent AC, Austin OL Jr, 1364–1392. Washington, DC: Smithsonian Institution Press; Houtman AM, Falls JB. 1994. Negative assortative mating in the white-throated sparrow, *Zonotrichia albicollis*: The role of mate choice and intra-sexual competition. *Animal Behaviour* 48:377–383.

3. Falls JB, Kopachena JG. 2010. White-throated sparrow (*Zonotrichia albicollis*), version 2.0. In *The Birds of North America*, edited by Poole AF. Ithaca, NY: Cornell Lab of Ornithology. https://doi.org/10.2173/bna.128.

Page 114. Colonial Bonds: house sparrow (*Passer domesticus*)

1. Cooper CB, Hochachka WM, Dhondt AA. 2007. Contrasting natural experiments confirm competition between house finches and house sparrows. *Ecology* 88:864–870.

2. Summers-Smith D. 1958. Nest-site selection, pair formation and territory in the house-sparrow *Passer domesticus*. *Ibis* 100:190-205; Summers-Smith D. 1956. Mortality of the house sparrow. *Bird Study* 3:265–270.

3. Cordero PJ, Wetton JH, Parkin DT. 1999. Extra-pair paternity and male badge size in the house sparrow. *Journal of Avian Biology* 30:97–102.

4. Summers-Smith J. 2003 The decline of the House Sparrow: a review. *British Birds* 96:439–446.

5 Cheptou P-O, Avendaño V LG. 2006. Pollination processes and the Allee effect in highly fragmented populations: consequences for the mating system in urban environments. *New Phytologist* 172:774–783.

6. Allee WC. 1931. *Animal Aggregations, a Study in General Sociology*. Chicago: University of Chicago Press.

Page 115. Rape: Canada goose (*Branta canadensis*)

1. Raveling DG. 1988. Mate retention of giant Canada geese. *Canadian Journal of Zoology* 66:2766–2768.

2. Moore JA, Kamarainen AM, Scribner KT, Mykut C, Prince HH. 2012. The effects of anthropogenic alteration of nesting habitat on rates of extra-pair fertilization and intraspecific brood parasitism in Canada geese *Branta canadensis*. *Ibis* 154:354–362.

3. Griffith SC, Owens IPF, Thuman KA. 2002. Extra pair paternity in birds: a review of interspecific variation and adaptive function. *Molecular Ecology* 11:2195–2212.

4. Whitford PC. 1993. Observations of attempted rape (forced copulation) in Canada geese. *Passenger Pigeon* 55:359–361.

5. McKinney F, Derrickson SR, Mineau P. 1983. Forced copulation in waterfowl. *Behaviour* 86:250–293.

6. Adler M. 2010. Sexual conflict in waterfowl: Why do females resist extrapair copulations? *Behavioral Ecology* 21:182–192.

Page 116. Domestication of Sex: Canada goose (*Branta canadensis*)

1. Wilson A. 1814. Canada goose. In *American Ornithology, or, The Natural History of the Birds of the United States,* vol. 8, 53–59. Philadelphia: Bradford and Inskeep.

2. Surrendi DC. 1970. The mortality, behavior, and homing of transplanted juvenile Canada geese. *The Journal of Wildlife Management* 34:719–733.

3. Bryant HC, Bailey FM, Bent AC, Christy BH, Preble EA, Saunders WE. 1936. Report of the Committee on Bird Protection American Ornithologists' Union. *The Auk* 53:70–73.

4. Nelson HK. 1963. Restoration of breeding Canada goose flocks in the north central states. In *Transactions of the Twenty-eighth North American Wildlife and Natural Resources Conference*, edited by Trefethen JB, 133–150. Detroit, MI: Wildlife Management Institute.

5. Hanson HC. 1997. *The Giant Canada Goose*, rev. ed. Carbondale and Edwardsville: Southern Illinois University Press.

6. Nelson HK, Oetting R. 1998. Giant Canada goose flocks in the United States. In *Biology and Management of Canada Geese.* Proceedings of the International Canada Goose Symposium, Held in Milwaukee, Wisconsin, 23-25 April 1991, edited by Rusch SH, Samuel MD, Humburg DD, Sullivan BD, 483–495. Milwaukee: International Canada Goose Symposium.

7. Conover MR. 2011. Population growth and movements of Canada geese in New Haven County, Connecticut, during a 25-year period. *Waterbirds* 34:412–421.

8. Malecki RA, Batt BDJ, Sheaffer SE. 2001. Spatial and temporal distribution of Atlantic population Canada geese. *The Journal of Wildlife Management* 65:242–247.

Page 117. Brood Parasite: brown-headed cowbird (*Molothrus ater*)

1. Burhans DE, Thompson FR. 2006. Songbird abundance and parasitism differ between urban and rural shrublands. *Ecological Applications* 16:394–405.

2. Peer BD, Sealy SG. 2004. Correlates of egg rejection in hosts of the brown-headed cowbird. *The Condor* 106:580–599.

3. Rasmussen JL, Sealy SG, Underwood TJ. 2009. Video recording reveals the method of ejection of brown-headed cowbird eggs and no cost in American robins and gray catbirds. *The Condor* 111:570–574.

4. Rothstein SI. 1982. Mechanisms of avian egg recognition: Which egg parameters elicit responses by rejecter species? *Behavioral Ecology and Sociobiology* 11:229-239.

5. Morrison ML, Hahn DC. 2002. Geographic variation in cowbird distribution, abundance, and parasitism. *Studies in Avian Biology* 25:65–72.

6. Swan DC. 2018. *Nest Predation by Brown-Headed Cowbirds (*Molothrus ater*).* Dissertation. London, Ontario: The University of Western Ontario.

7. Hoover J, Robinson S. 2007. Retaliatory mafia behavior by a parasitic cowbird favors host acceptance of parasitic eggs. *Proceedings of the National Academy of Sciences* 104:4479–4483.

8. Swan D, Zanette L, Clinchy M. 2015. Brood parasites manipulate their hosts: Experimental evidence for the farming hypothesis. *Animal Behavior* 105:29–35.

9. Peer BD, Rivers JW, Merrill L, Robinson SK, Rothstein SI. 2017. The brown-headed cowbird: A model species for testing novel research questions in animal ecology, evolution, and behavior. In *Avian Brood Parasitism Behaviour, Ecology, Evolution and Coevolution*, edited by Soler M, 161–187. Cham, Switzerland: Springer.

Page 118. Polygyny: European starling (*Sturnus vulgaris*)

1. Kessel B. 1957. A study of the breeding biology of the European starling (*Sturnus vulgaris* L.) in North America. *The American Midland Naturalist* 58:257–331.

2. Pinxten R, Eens M, Verheyen RF. 1989. Polygyny in the European starling. *Behaviour* 111:234–256.

3. Smith HG, Sandell MI. 2005. The starling mating system as an outcome of the sexual conflict. *Evolutionary Ecology Research* 19:151–165.

4. Eens M, Pinxten R. 1995. Mate desertion by primary female European starlings at the end of the nestling stage. *Journal of Avian Biology* 26:267-271.

5. Mennechez G, Clergeau P. 2006. Effect of urbanisation on habitat generalists: starlings not so flexible? *Acta Oecologica* 30:182–191.

Page 119. Chapter 13 Fungi, Bacteria, and Lichens

1. Seminara A, Fritz J, Brenner MP, Pringle A. 2018. A universal growth limit for circular lichens. *Journal of the Royal Society Interface* 15:20180063. https://doi.org/10.1098/rsif.2018.0063.

Page 120. Microbial Mating: bacteria

1. Xu J. 2004. The prevalence and evolution of sex in microorganisms. *Genome* 47:775–780.

2. Broszat M, Grohmann E. 2014. Horizontal gene transfer in planktonic and biofilm modes. In *Antibiofilm Agents*, edited by Rumbaugh KP, Ahmad I, 67–95. Berlin-Heidelberg: Springer-Verlag.

3. Redfield RJ. 2012. Do bacteria have sex? In *Microbes and Evolution: The World That Darwin Never Saw*, edited by Kolter R, Maloy S, 136–143. Washington, DC: ASM Press.

4. Cavalier-Smith T. 2002. Origins of the machinery of recombination and sex. *Heredity* 88:125–141.

5. Bryers JD. 1987. Biologically active surfaces: Processes governing the formation and persistence of biofilms. *Biotechnology Progress* 3:57–68.

6. Watnick P, Kolter R. 2000. Biofilm, city of microbes. *Journal of Bacteriology* 182:2675–2679.

7. Hausner M, Wuertz S. 1999. High rates of conjugation in bacterial biofilms as determined by quantitative in situ analysis. *Applied and Environmental Microbiology* 65:3710 3713.

8. Clewell DB. 1993. Bacterial sex pheromone–induced plasmid transfer. *Cell* 73:9–12.

9. Flemming H-C, Wingender J. 2010. The biofilm matrix. *Nature Reviews Microbiology* 8:623–633.

10. van Schaik W. 2015. The human gut resistome. *Philosophical Transactions of the Royal Society B* 370:20140087. https://doi.org/10.1098/rstb.2014.0087.

11. Marcinek H, Wirth R, Muscholl-Silberhorn A, Gauer M. 1998. *Enterococcus faecalis* gene transfer under natural conditions in municipal sewage water treatment plants. *Applied and Environmental Microbiology* 64:626–632.

12. Crispim CA, Gaylarde PM, Gaylarde CC. 2003. Algal and cyanobacterial biofilms on calcareous historic buildings. *Current Microbiology* 46:79–82.

Page 121. Vegetative Reproduction: lemon lichen (*Candelaria concolor*)

1. Hawksworth DL. 1988. The variety of fungal-algal symbioses, their evolutionary significance, and the nature of lichens. *Journal of the Linnean Society* 98:3–20.

2. Murtagh G, Dyer P, Crittenden P. 2000. Sex and the single lichen. *Nature* 404:564.

3. Garty J, Delarea J. 1988. Evidence of liberation of lichen ascospores in clusters and reports on contact between free-living algal cells and germinating lichen ascospores under natural conditions. *Canadian Journal of Botany* 66:2171–2177.

4. Murtagh et al. Sex and the single lichen.

5. Tripp EA, Lendemer JC. 2018. Twenty-seven modes of reproduction in the obligate lichen symbiosis. *Brittonia* 70:1–14. https://doi.org/10.1007/s12228-017-9500-6.

6. McDonald L, Woudenberg MV, Dorin B, Adcock AM, McMullin RT, Cottenie K. 2017. The effects of bark quality on corticolous lichen community composition in urban parks of southern Ontario. *Botany* 95:1141–1149; Hyerczyk RD. 2005. The lichen flora of ten Chicago parks: Chicago Park District, Chica-

go, Illinois. *Transactions of the Illinois State Academy of Science* 98:97–122.

7. Brodo IM, Sharnoff SD, Sharnoff S. 2001. *Lichens of North America*. New Haven, CT: Yale University Press.

8 . Tripp EA. 2016. Is asexual reproduction an evolutionary dead end in lichens? *The Lichenologist* 48:559–580.

Page 122. Mating Types: mushroom-forming fungi (**Agaricomycetes**)

1. Kües U. 2000. Life history and developmental processes in the basidiomycete *Coprinus cinereus. Microbiology and Molecular Biology Reviews* 64:316–353.

2. May G, Taylor JW. 1988. Patterns of mating and mitochondrial DNA inheritance in the agaric basidiomycete *Coprinus cinereus. Genetics* 118:213–220.

3. Vreeburg S, Nygren K, Aanen DK. 2016. Unholy marriages and eternal triangles: How competition in the mushroom life cycle can lead to genomic conflict. *Philosophical Transactions of the Royal Society B* 371:20150533. https://doi.org/10.1098/rstb.2015.0533.

4. James TY. 2015. Why mushrooms have evolved to be so promiscuous: Insights from evolutionary and ecological patterns. *Fungal Biology Reviews* 29:167–178.

5. Brown AJ, Casselton LA. 2001. Mating in mushrooms: Increasing the chances but prolonging the affair. *Trends in Genetics* 17:393–400.

6. James, Why mushrooms have evolved to be so promiscuous.

Page 123. Microbial-Induced Change in Sex: pill bug (*Armadillidium vulgare*)

1. Rigaud T, Juchault P, Mocquard J-P. 1997. The evolution of sex determination in isopod crustaceans. *BioEssays* 19:409–416.

2. Wilson EO. 1988. The current state of biological diversity. In *Biodiversity*, edited by Wilson EO, Peter FM, 3–18. Washington, DC: The National Academies Press.

3. Weinert LA, Araujo-Jnr EV, Ahmed MZ, Welch JJ. 2015. The incidence of bacterial endosymbionts in terrestrial arthropods. *Proceedings of the Royal Society B* 282:20150249. https://doi.org/10.1098/rspb.2015.0249.

4. Werren JH, Baldo L, Clark ME. 2008. *Wolbachia*: Master manipulators of invertebrate biology. *Nature Reviews Microbiology* 6:741–751.

Page 124. Sex for Biological Control: Asian tiger mosquito (*Aedes albopictus*)

1. Faraji A, Unlu I. 2016. The eye of the tiger, the thrill of the fight: Effective larval and adult control measures against the Asian tiger mosquito, *Aedes albopictus* (Diptera: Culicidae), in North America. *Journal of Medical Entomology* 53:1029–1047.

2. Hertig M, Wolbach SB. 1924. Studies on *Rickettsia*-like micro-organisms in insects. *The Journal of Medical Research* 44:329–374, plates XXVII–XXX; Hertig M. 1936. The Rickettsia, *Wolbachia pipientis* (gen. et sp.n.) and associated inclusions of the mosquito, *Culex pipiens. Parasitology* 28:453–486.

3. Calvitti M, Moretti R, Skidmore AR, Dobson SL. 2012. *Wolbachia* strain wPip yields a pattern of cytoplasmic incompatibility enhancing a *Wolbachia*-based suppression strategy against the disease vector *Aedes albopictus. Parasites & Vectors* 5:254. https://doi.org/10.1186/1756-3305-5-254.

4. Crawford JE, Clarke DW, Criswell V, Desnoyer M, et al. 2020. Efficient production of male *Wolbachia*-infected *Aedes aegypti* mosquitoes enables large-scale suppression of wild populations. *Nature Biotechnology* 38:482–492; Mains JW, Kelly PH, Dobson KL, Petrie WD, Dobson SL. 2019. Localized control of *Aedes aegypti* (Diptera: Culicidae) in Miami, FL, via inundative releases of *Wolbachia*-infected male mosquitoes. *Journal of Economic Entomology* 56:1296–1303; Zheng X, Zhang D, Li Y, Yang C, et al. 2019. Incompatible and sterile insect techniques combined eliminate mosquitoes. *Nature* 572:56–61.

5. Ant T, Herd C, Geoghegan V, Hoffmann A, Sinkins S. 2018. The *Wolbachia* strain wAu provides highly efficient virus transmission blocking in *Aedes aegypti. PLOS Pathogens* 14:e1006815. https://doi.org/10.1371/journal.ppat.1006815; Mancini MV, Herd CS, Ant TH, Murdochy SM, Sinkins SP. 2020. *Wolbachia* strain wAu efficiently blocks arbovirus transmission in *Aedes albopictus. PLOS Neglected Tropical Diseases* 14:e0007926. https://doi.org/10.1371/journal.pntd.0007926.

6. Ferreira AG, Fairlie S, Moreira LA. 2020. Insect vectors endosymbionts as solutions against diseases. *Current Opinion in Insect Science* 40:56–61.

7. Li M, et al. 2020. Development of a confinable gene drive system in the human disease vector *Aedes aegypti. eLife* 9:e51701. https://doi.org/10.7554/eLife.51701.

Page 125. Conclusion

1. Bartlewicz J, Vandepitte K, Jacquemyn H, Honnay O. 2015. Population genetic diversity of the clonal self-incompatible herbaceous plant *Linaria vulgaris* along an urbanization gradient. *Biological Journal of the Linnean Society* 116:603–613; Gorton AJ, Moeller DA, Tiffin P. 2018. Little plant, big city: A test of adaptation to urban environments in common ragweed (*Ambrosia artemisiifolia*). *Proceedings of the Royal Society B* 285:20180968. https://doi.org/10.1098/rspb.2018.0968.

2. Desaegher J, Nadot S, Machon N, Colas B. 2019. How does urbanization affect the reproductive characteristics and ecological affinities of street plant communities? *Ecology and Evolution* 9:9977–9989.

3. Aguilar R, Quesada M, Ashworth L, Herrerias-Diego Y, Lobo J. 2008. Genetic consequences of habitat fragmentation in plant populations: Susceptible signals in plant traits and methodological approaches. *Molecular Ecology* 17:5177–5188.

4. Biesmeijer JC, et al. 2006. Parallel declines in pollinators and insect-pollinated plants in Britain and the Netherlands. *Science* 313:351–354.

5. Goodwillie C, Sargent RD, Eckert CG, Elle E, et al. 2010. Correlated evolution of mating system and floral display traits in flowering plants and its implications for the distribution of mating system variation. *New Phytologist* 185:311–321.

6. Snell R, Aarssen LW. 2005. Life history traits in selfing versus outcrossing annuals: Exploring the 'time-limitation' hypothesis for the fitness benefit of self-pollination. *BMC Ecology* 5. https://doi.org/10.1186/1472-6785-5-2.

7. Costea M, Tardif FJ. 2005. Biology of Canadian Weeds: 131. *Polygonum aviculare* L. *Canadian Journal of Plant Science* 85:481-506.; Bogle AL. 1970. The genera of Molluginaceae and Aizoaceae in the southeastern United States. *Journal of the Arnold Arboretum* 51:43–462.

8. Cheptou P-O. 2019. Does the evolution of self-fertilization rescue populations or increase the risk of extinction? *Annals of Botany* 123:337–345.

Page 126. Glossary

1. Asker S, Jerling L. 1992. *Apomixis in Plants.* Boca Raton, FL: CRC Press; Grusz AL. 2016. A current perspective on apomixis in ferns. *Journal of Systematics and Evolution* 54: 656–665.

2. Wilschut RA, Oplaat C, Snoek LB, Kirschner J, Verhoeven KJF. 2016. Natural epigenetic variation contributes to heritable flowering divergence in a widespread asexual dandelion lineage. *Molecular Ecology* 25:1759–1768.

3. Gorton AJ, Moeller DA, Tiffin P. 2018. Little plant, big city: a test of adaptation to urban environments in common ragweed (*Ambrosia artemisiifolia*). *Proceedings of the Royal Society B* 285:20180968. https://doi.org/10.1098/rspb.2018.0968

4. Hitchings SP, Beebee TJC. 1998. Loss of genetic diversity and fitness in common toad (*Bufo bufo*) populations isolated by inimical habitat. *Journal of Evolutionary Biology* 11:269–283.

5. Munshi-South J, Zak Y, Pehek E. 2013. Conservation genetics of extremely isolated urban populations of the northern dusky salamander (*Desmognathus fuscus*) in New York City. *Peer J* 1:e64 https://doi.org/10.7717/peerj.64.

6. Harrison JS, Mondor EB. 2011. Evidence for an invasive aphid "superclone." Extremely low genetic diversity in oleander aphid (*Aphis nerii*) populations in the southern United States. *PloS One* 6:e17524. https//doi.org/10.1371/journal.pone.0017524.

7. Schlupp I. 2005. The evolutionary ecology of gynogenesis. *Annual Review of Ecology and Systematics* 36:399–417.

8. Heimpel GE, de Boer JG. 2008. Sex determination in the Hymenoptera. *Annual Review of Entomology* 53:209–230.

9. Darwin C. 1876. *The Effects of Cross and Self Fertilisation in the Vegetable Kingdom.* London: John Murray.

10. de Carvalho JF, Oplaat C, Pappas N, Derks M, Ridder Dd, Verhoeven KJF. 2016. Heritable gene expression differences between apomictic clone members in *Taraxacum officinale*: Insights into early stages of evolutionary divergence in asexual plants. *BMC Genomics* 17:203. https://doi.org/10.1186/s12864-016-2524-6.

11. Jaenike J, Selander R. 1979. Evolution and ecology of parthenogenesis in earthworms. *American Zoologist* 19:729–737.

12. Hohenlohe PA, Arnold SJ. 2010. Dimensionality of mate choice, sexual isolation, and speciation. *Pro-*

ceedings of the National Academy of Sciences 107:16583–16588.

13. Klekowski EJ. 1973. Sexual and subsexual systems in homosporous pteridophytes: a new hypothesis. *American Journal of Botany* 60:535–544.

14. van Baarlen P, van Dijk P, Hoekstra R, de Jong J. 2000. Meiotic recombination in sexual diploid and apomictic triploid dandelions (*Taraxacum officinale* L.). *Genome* 43:827–835.

15. Tryon AF, Britton DM. 1958. Cytotaxonomic studies on the fern genus *Pellaea*. *Evolution* 12:137–145.

16. Iaffaldano B, Zhang Y, Cardina J, Cornis K. 2017. Genome size variation among common dandelion accessions informs their mode of reproduction and suggests the absence of sexual diploids in North America. *Plant Systematics and Evolution* 303:719–725.

17. Lee T, Siripattrawan S, Ituarte CF, Foighil DÓ. 2005. Invasion of the clonal clams: *Corbicula* lineages in the New World. *American Malacological Bulletin* 20:113–122.

Photo Credits

Page

/ INDEX /

Berkeley, California, 106
Berlin, Germany, xxviii
billing, 107, 109
bills, adaptations in, 104
biodiversity, xxiii
biofilms, 119, 120
biological control of mosquitoes and their
    viruses, 124
Bipalium (terrestrial flatworms), 84
    adventitium, 84
birds, 101–18. *See also specific bird names*
    bill adaptations in, 104
    brood parasitism in, 115, 117
    colonial bonds in, 114
    conditional sexual traits in, 103
    courtship in, 101, 105, 107–10
    domestication of sex in, 116
    extra-pair paternity in, 115
    forced copulation in, 115
    insect defenses against, 46
    noise and, xvi, xxix, 111
    northern house mosquitoes and, 70
    polygyny in, 118
    predation by, 44, 87
    repetitive copulation in, 106
    seed dispersal and, 10, 27
    singing in, 111, 112, 113
    submissive males in, 104
    territoriality in, 102, 112
    vacant lots and, xvi
bisexual flowers, xiii, 10, 11, 13, 14, 23, 32
bisexual plants, 2, 35
bittersweets (*Celastrus*), 37
Bitzer, Royce, 43
black locust (*Robinia pseudoacacia*), 68
Blatta orientalis (oriental cockroach), xvi, 65,
    71
Blattella germanica (German cockroach), 71
blue dasher dragonfly (*Pachydiplax
    longipennis*), 58–60, 62, 96
bluegill (*Lepomis macrochirus*), 67, 90, 92–94
bluegrass, annual (*Poa annua*), xv, xx, 10, 31
blue violet, common (*Viola sororia*), 14
Bombus
    citrinus (lemon cuckoo bumblebee), 52
    impatiens (common eastern bumblebee), 21,
        51–52
boneset, late (*Eupatorium serotinum*), 48
boneset, tall (*Eupatorium altissimum*), 68
boreal forests, 113
Borge, Mary Anne, 51
Boston, Massachusetts, 70
box elder (*Acer negundo*), 34
brackish water organisms, ballast and
    introduction of, xvii
Branta canadensis (Canada goose), 101,
    115–16

Brazil, 8
bridge spider (*Larinioides sclopetarius*), 75,
    76–78, 98
British robin (*Erithacus rubecula*), 105
brood parasitism, 115, 117
broods, repeated, 106
brown-headed cowbird (*Molothrus ater*), 101,
    117
Bruce, Hilda, 100
Bruder, Kenneth, 56
Bryum argenteum (silvergreen bryum moss), 8
Buddleia (butterfly bush), 63
Bufo bufo (common toad), 90
bullfrog, American (*Lithobates catesbeianus*),
    85, 89
bumblebees, 51–52
    anti-aphrodisiac pheromone in, 51, 86
    female usurpers and, 52
    mating plugs in, 51, 97
    matricide in, 50
    nectar robbery in jewelweed by, 12
    pollination by, 12, 13, 26, 27
    polyandry in, 51
    white campion and, 21
Buteo jamaicensis (red-tailed hawk), 101, 102,
    105
butterflies, 39, 40–45. *See also specific
    butterfly names*
butterfly bush (*Buddleia*), 63

cabbage white butterfly (*Pieris rapae*), 13, 39,
    40–41, 86, 97
caddisflies (Trichoptera), 61
Calvitti, Maurizio, 124
calyx tube, fused, 21
Canada goose (*Branta canadensis*), 101,
    115–16
Candelaria concolor (lemon lichen), 121
cannibalism, 62, 63–64, 66, 75, 78, 92
cardinal, northern (*Cardinalis cardinalis*), 101
Carolina mantis (*Stagmomantis carolina*), 63,
    64
carotenoids, 103
carpetweed (*Mollugo verticillata*), 9, 125
cecropia moth (*Hyalophora cecropia*), 39, 47
Celastrus (bittersweets), 37
    orbiculatus (Oriental bittersweet), 37
    scandens (American bittersweet), 37
cemeteries, xvii, xxiii, 61, 119
centipedes, xvi, 80
Chaitophorus populicola (poplar leaf aphid),
    73
Cheiracanthium (yellow sac spider), 79
    mildei, 79
Cheng, Mei-Fang, 106, 111
Chicago, Illinois, 70, 99

Linaria vulgaris (yellow toadflax), 26, 114, 125

Lithobates catesbeianus (American bullfrog), 85, 89

Lithobates palustris (pickerel frog), 89

liverworts, xv, 3–5, 15

llamas, 99

locust, black (*Robinia pseudoacacia*), 68

locust borer (*Megacyllene robiniae*), 65, 68

London, England, 4, 70, 114

London plane trees (*Platanus × acerifolia*), 33

long-distance sex pheromones, 47, 64

Longley, Ariana, 26

looper moth, common (*Autographa precationis*), xvii, 21

lost sex, 7, 32

Lucanus capreolus (reddish-brown stag beetle), 67

Lumbricus terrestris (nightcrawler earthworm), 83, 84

luteinizing (sex) hormone, 111

"mafia" hypothesis for nest predation, 117

male display in courtship, 107, 108–9

male dominance, 38

male genitalia, two sets of, 77

male interference, 37

male-male competition, 16, 67, 69

male mimicry of females, 87, 93

male offspring from unfertilized eggs, 50, 55

males, submissive, 104

male stuffing, 53

male territoriality, 43, 58, 61

male-to-female sex change in flowers, 13

mallard ducks, 116

mammals, xvi, 70, 83, 95–100

mandibles, giant, 67

Mantidae, 63–64

mantises, 57, 63–64

Mantis religiosa (European mantis), 57

maple trees, 34–36

marble screw-moss (*Syntrichia papillosa*), 4, 7

Marchantia
  inflexa, 4
  polymorpha (umbrella liverwort), xv, 3–4, 5, 15

Maron, John, 25

masking of bird song, 111

mating. *See also* copulation
  alternative tactics for, 54, 90, 98
  amplexus in toads, 89, 90
  cannibalism and, 63
  efficiency of, 28
  female preferences for, 40, 42, 44, 67, 89, 98, 103
  infertile, 48
  injury caused during, 74, 83
  light pollution and, xvi, xxvi, 21
  microbial, 120
  mixed modes of, 25
  vulnerable places for, 68
  without copulation, 80

mating aggregations, 86, 87, 91

mating balls, 89

mating calls, xxix, 85, 89, 95

mating chases, 98

mating plugs, 51, 86, 97

mating systems
  of European starlings, 118
  of herbaceous annuals, 9
  of herbaceous perennials, 19
  of invertebrates, 75
  mixed, 25
  of moths and butterflies, 39
  self-incompatible mating types and, 26
  of urban woody plants, 33

mating trails, lethal, 84

mating types, 26, 122

matriarchy, 52

matricide, 50, 52

Matzk, Fritz, 25

mauling of worker bees, 52

mayflies (Ephemeroptera), 61

Megacyllene robiniae (locust borer), 65, 68

messenger RNA (mRNA), 18, 20

Metepeira labyrinthea (labyrinth orb weaver), 78

metolachlor, 94

mice, xvi, 95, 100

microbial-induced sex change, 123

microbial mating, 120

Microbotryum lychnidis-dioicae (anther smut fungus), 20

Micropterus (bass), 94

microRNAs, 18

midges, 76

Miles, Lindsay, 79

milkweed, common (*Asclepias syriaca*), 19, 28, 39, 69, 72

milkweed aphid (*Aphis nerii*), 72, 73

millipedes, xvi, 80

mimetic sex pheromone, 87

mimicry, sexual, 44–45, 57, 60, 87, 93

Minneapolis, Minnesota, 17

Mission Dolores Park, San Francisco, xxii

mites, 80

mitochondrial DNA, 81

Mollugo verticillata (carpetweed), 9, 125

Molothrus ater (brown-headed cowbird), 101, 117

monandry, 62

Monfalcone, Italy, xix

monogamy, 50, 51. *See also* social monogamy

Montpellier, France, 114
Morus (mulberries), 29, 38
  alba (white mulberry), 38
  rubra (red mulberry), 38
Moscow, Idaho, 31
mosquitoes, 65, 70, 124
mosses, 1, 2, 4, 7, 8
"moth night" at Bartram's Garden, xxvii
moths, 39, 46–48
  eaten by house centipedes, 80
  infertile mating in, 48
  long-distance sex pheromone in, 47
  nocturnal light pollution and, 21
  sequestered females, 46
  traumatic insemination in, 74
mourning dove (*Zenaida macroura*), 101, 105,
  106–11
mugwort (*Artemisia vulgaris*), 18, 30, 68
mulberries (*Morus*), 29, 38
Munshi-South, Jason, xvi
mushroom-forming fungi (Agaricomycetes),
  119, 122
Mus musculus (house mouse), 95, 100
mycelia, 122

Nazareth, Israel, xxi
neckweed (*Veronica peregrina*), 14
nectar theft/robbery, 12, 28, 39
nest predation, 117
nests, abandonment of, 103, 118
New York City, xvi, xvii, xviii, 37, 38, 70
nightcrawler earthworm (*Lumbricus
  terrestris*), 83, 84
nocturnal fertilization, 21
nocturnal light pollution, xvii, 21, 46, 48, 76,
  77, 78
noise pollution, xvi, xxix, 85, 89, 111, 112
nonflowering plants, 1–8
Norfolk, England, 8
northern cardinal (*Cardinalis cardinalis*), 101
northern goshawk (*Accipiter gentilis*), 106
northern house mosquito (*Culex pipiens*), 65,
  70, 124
Norway maple (*Acer platanoides*), 36
Norway rats, xvi
nuclei, fusion of, 122
nuptial dance, 82
nuptial gifts, 82
nursing and lactation, 95, 100

ocelli (simple eyes), 58, 63
Octolasion tyrtaeum (earthworm), 83
Oden, Neal, 35
Odonata, 57, 61. *See also* damselflies;
  dragonflies

oleander aphids. *See* milkweed aphid (*Aphis
  nerii*)
open *vs.* closed types of flowers, 14
orange sulfur butterfly (*Colias eurytheme*), 42,
  67
orchard oriole (*Icterus spurius*), 117
orchids, 28
organochlorine pesticides, 94
Oriental bittersweet (*Celastrus orbiculatus*),
  37
oriental cockroach (*Blatta orientalis*), xvi, 65,
  71
oriole, orchard (*Icterus spurius*), 117
Orthotrichum pumilum (dwarf bristle-moss), 2
Oryctolagus cuniculus (European rabbit), 99
outbreeding, xv, 12, 100
outcrossing
  gardens and, xvi
  in jewelweed, 13
  lawns and vacant lots and, xvii
  in mushroom-forming fungi, 122
  in ragweed, 16
  segregation of sexes and, 34
  self-incompatibility mating system and, 26
  urbanization and decrease in, 26
ovulation, copulation-induced, 99
Ownbey, Marion, 31
Oxalis corniculata (creeping wood sorrel), 22

Pachydiplax longipennis (blue dasher
  dragonfly), 58–60, 62, 96
painted lady butterfly (*Vanessa cardui*), 63
panmixia, 91
paper wasps, 53–54, 55, 70, 90
Papilio glaucus (eastern tiger swallowtail),
  44–45
parasites, 20, 52, 99
parasitic infections and polyandry, 51
parasitic vines, flowering, 18, 20
parasitism, brood, 115, 117
parasitism, evolutionary arms races and, 18, 20
parental investment, equal, 55
parenting alone, 92
Paris, France, xxx, 26, 125
parthenogenesis, 71–73, 83, 84
passenger pigeon, 114
Passer domesticus (house sparrow), 114
paternal care, 103
paternal inheritance, 81
pathogens, 99
pavement ant (*Tetramorium immigrans*), 56
pavement cracks. *See* sidewalk and pavement
  cracks
pecking orders, 104
pedestrian traffic, xv, 10, 27
pedipalps, 77

in American toads, 80, 89, 90
  in bluegills, 92, 93
  noise and, 111
spermatophores, 40
sperm competition for fertilization, 41, 83, 86
sperm containers or packages, 40, 80
sperm inadequacy, 48
sperm platforms of umbrella liverwort, 3
sperm transport in water, 3, 5, 6
Sphecius speciosus (cicada killer), 54, 55
spiders, 76–79
  as arthropods, 123
  house, 79
  injury to mates in, 83
  loss of aggressiveness in, 78, 98
  mating promoted by nocturnal light
      pollution, xvi, 76, 77
  in the soil, xvi
  vibratory noise and, 111
splash cups of umbrella liverwort, 4
spleenwort, ebony (*Asplenia platyneuron*), 6
sporadic sex, 8
spore capsules in nonflowering plants, xiv, 2,
      3, 4, 5, 8
spore dispersal, xxiv, 2, 4, 5, 6
spore-producing structures, 1, 121
spores, by self-fertilization, 2, 5
spores without fertilization, 6
springtails, xvi, 80
squid, giant, 74
squirrel, eastern gray (*Sciurus carolinensis*),
      42, 67, 90, 95, 96–98
stabilization in dragonflies, 58
stag beetles, 67
Stagmomantis carolina (Carolina mantis), 63,
      64
starling, European (*Sturnus vulgaris*), 118
Stebbins, George Ledyard, 23
Stellaria media (common chickweed), 5,
      10–11, 13, 25, 31, 125
sterilization, *Wolbachia*-induced, 124
Sternburg, James, 47
stigmas, defined, 10, 16
St. John's wort (*Hypericum perforatum*), 25
stoneflies (Plecoptera), 61
St. Peters Church graveyard, 119
Sturnus vulgaris (European starling), 118
submissive males, 104
Summers-Smith, James Denis, 114
superclones, 30
swallowtail butterflies, 44–45
swamp darner (*Epiaeschna heros*), 61
Switzerland, 21
sycamores, 33
Sylvilagus floridanus (eastern cottontail), 95,
      99
symbiotic partnerships, 119, 121, 123

Syntrichia papillosa (marble screw-moss), 4, 7
syrphids (flower flies), 54

Takeuchi, Tsuyoshi, 43
tall boneset (*Eupatorium altissimum*), 68
Taraxacum officinale (common dandelion),
      23, 25
Taylor, Orley R., 48
telescoping of generations, 72
temperature-dependent sex determination, 88
Tenodera sinensis (Chinese mantis), 63
terrestrial flatworms (*Bipalium*), 84
territoriality, 43, 58, 61, 96, 100, 102, 112
testosterone, 103
Tetramorium immigrans (pavement ant), 56
Tetraopes tetraophthalmus (red milkweed
      beetle), 69
Texas A&M University, 106
Thamnophis sirtalis (common garter snake),
      13, 86–87, 97
thermoregulation, 58
threadstalk speedwell (*Veronica filiformis*), xi,
      7, 30, 32
thrips, 10, 11, 28
Thyridopteryx ephemeraeformis (evergreen
      bagworm moth), 39, 46
tiger swallowtail, eastern (*Papilio glaucus*),
      44–45
toads, 42, 80, 89–90
Toh, Su San, 20
Tokyo, Japan, xxvi
Trachemys scripta elegans (red-eared slider
      turtle), 88
Tragopogon (salsifies), 31
  dubius (yellow salsify), 31
  mirus, 31
  miscellus, 31
  porrifolius, 31
  pratensis, 31
transportation and species spread, 29, 30
traumatic insemination, 74
tree of heaven (*Ailanthus altissima*), 48
trees, 33, 34–36, 38. *See also specific trees*
Treptowers, Berlin, xxviii
triazines, 94
Trichoptera (caddisflies), 61
Trifolium repens (white clover), xvii, 27
Trinidad, 4
Triodanis perfoliata (Venus's looking glass),
      14
true bugs (Hemiptera), 61
trypanosomes, 51
Turdus migratorius (American robin), 112, 117
turf, 23, 27. *See also* lawns and vacant lots
turtle, red-eared slider (*Trachemys scripta
      elegans*), 88

tympanum, 85

ultrasonic courtship sounds, 95
umbrella liverwort (*Marchantia polymorpha*), xv, 3–4, 5, 15
unfertilized eggs and male offspring, 50, 55
unisexual colonies, 4, 7, 71
unisexual function in bisexual flowers, 10, 11, 13, 14
urban canyons, xvi, xviii
urban forests, xvi, xvii
urbanization
  genetic diversity and, 26, 27, 51
  genetic isolation and, 86
  herbaceous annuals and, 9, 17
  herbaceous perennials and, 19, 26, 27, 29
  lichens and, 121
  polygyny and, 118
  sexual isolation and, 17
  shift to self-fertilization and, 125

vacant lots and lawns, xvii, xxi, xxii, 27, 29, 32
Valley Parkway, Ohio, xxv
Vanessa
  atalanta (red admiral butterfly), 43
  cardui (painted lady butterfly), 63
vegetative propagation
  in lemon lichen, 121
  in mugwort, 18, 30
  in nonflowering plants, 4, 5, 6, 7, 8
  in threadstalk speedwell, 32
vehicular traffic
  insect defenses against, 46
  noise and, xvi, 85, 89, 111
  seed dispersal and, xv, xxiv, 10, 27
  spore dispersal and, 2
Venus's looking glass (*Triodanis perfoliata*), 14
Verboven, Hans, 27
Veronica
  filiformis (threadstalk speedwell), xi, 7, 30, 32
  peregrina (neckweed), 14
Vespula maculifrons (eastern yellowjacket), 49
vines, 18, 37
Viola sororia (common blue violet), 14
viruses, 120, 124

Waldbauer, Gilbert, 47
warning positions, 58
Warsaw, Poland., xxiv
Washington, D.C., xxix
wasps, 28, 49, 50, 53–54, 55, 70, 90

waterfowl, 115
water pollution, xvi, xxx
webworm moth, ailanthus (*Atteva aurea*), 48
western black widow spider (*Latrodectus hesperus*), 78, 79
western scrub jay, 111
Weston, Leslie, 30
wetlands, 29
white campion (*Silene latifolia*), 15, 20–21
white clover (*Trifolium repens*), xvii, 27
white-footed mice, xvi
white mulberry (*Morus alba*), 38
white-throated sparrow (*Zonotrichia albicollis*), 101, 113
Whitten, Wesley, 100
Wildermuth, Hansruedi, 61
Wild Urban Plants of the Northeast (del Tredici), 30
Wilson, Alexander, 116
winter survival, 103
Wolbach, Simon, 124
Wolbachia, 123, 124
wood sorrel, creeping (*Oxalis corniculata*), 22
woody plants, urban, 33–38
worms, xvi, 74, 83–84

yellow fever mosquito (*Aedes aegypti*), 124
yellowjackets, 49, 50, 54, 55
yellow sac spider (*Cheiracanthium*), 79
yellow salsify (*Tragopogon dubius*), 31
yellow toadflax (*Linaria vulgaris*), 26, 114, 125

Zabulon skipper (*Poanes zabulon*), 39
Zenaida macroura (mourning dove), 101, 105, 106–11
Zika virus, 124
Zimmerman, Michael, 12
Zonotrichia albicollis (white-throated sparrow), 101, 113

GPSR Authorized Representative: Easy Access System Europe, Mustamäe tee 50, 10621 Tallinn, Estonia, gpsr.requests@easproject.com